GENETIC REVOLUTION

Genetic Revolution

SHAPING LIFE FOR TOMORROW

By D. S. HALACY, Jr.

HARPER & ROW, PUBLISHERS

NEW YORK
EVANSTON
SAN FRANCISCO
LONDON

1817

GENETIC REVOLUTION: SHAPING LIFE FOR TOMORROW. Copyright © 1974 by Daniel S. Halacy, Jr. All rights reserved. Printed in the United States of America. No part of this book may be used or reproduced in any manner whatsoever without written permission except in the case of brief quotations embodied in critical articles and reviews. For information address Harper & Row, Publishers, Inc., 10 East 53rd Street, New York, N.Y. 10022. Published simultaneously in Canada by Fitzhenry & Whiteside Limited, Toronto.

FIRST EDITION

Designed by Sidney Feinberg

Library of Congress Cataloging in Publication Data

Halacy, Daniel Stephen, 1919–
  Genetic revolution.
  Bibliography: p.
  1. Human genetics. 2. Human genetics—Social aspects. I. Title. [DNLM: 1. Genetic intervention—Popular works. 2. Genetics, Human—Popular works. QH431 H155g 1974]
QH431.H2426   573.2′1   73–4085
ISBN 0–06–011713–3

Contents

Introduction

We have come a long way—scientifically and technologically—since Prometheus snatched fire from the gods, but even the loosing of nuclear fire was a Promethean feat. Now, by taking hold of the workings of life's spark itself, human beings have moved from robbing the gods to playing at gods. Alone among living creatures, man is beginning to comprehend not only what makes himself tick but what *makes* him, as well. Having discerned the secret of life, we are at the point of being able to influence its processes. In consequence, we shape the human future.

Proper or not, the present study of mankind is man; as chemists and physicists once poked into inert matter's molecules, atoms, electrons, and nuclei, the biophysicist today pries ever deeper within the living cell, its nucleus, chromosomes, and now even the genes themselves. The touchstone of old would have transmuted only metals, made gold from lead, a magic that still eludes the physicist. The biological revolution is concerned not with the manipulation of lifeless atoms but of living entities—we have become tinkers not only of pots and pans but of men as well.

Early in the game, the genetic revolution was a paper tiger; the biophysicist only vaguely considered matters that science-fiction writers had been describing for decades. But some of the paperwork is finished, and the bioscientist has moved from his ivy tower down into the hard light of the lab. Pioneer geneticists like Thomas Hunt Morgan and Hermann J. Muller mapped the chromosomes of fruit flies, and dreamed of sperm banks to produce a crop of Shakespeares and Einsteins from the mothers of America. Now a new generation is doing bigger things and boldly considering others few scientists would have considered seriously a short time ago.

When Aldous Huxley published *Brave New World* in 1932, his baby hatchery was science fiction, its characters playing out their roles in a world several hundred years hence. But some of the genetic feats of this prophetic novel have already been matched. Biological engineering is no longer a fiction but grist for prestigious journals. Quite suddenly, population genetics is not mere statistical exercise but a social engineering challenge, with the requisite hardware in sight.

Genetic counseling is accepted procedure; parents are advised of the odds of conceiving defective children, or told whether those already conceived will suffer. Hundreds of diseases, including cancer, are considered to be genetic to some extent. Two different aspects of genetics bear on current efforts to battle sickle-cell anemia —it is hereditary, but in addition it affects blacks almost exclusively. Other diseases single out different segments of the population.

Beyond counseling comes genetic therapy. Doctors now speak of the "information of cancer" or whatever other genetic disease, for the gene is indeed a molecule of information coding amino acids. When corrective information is implanted into a cell, disease can be thwarted, as demonstrated by inoculations of virus against many diseases. Proponents of gene therapy claim we will thus doctor the genes for a great variety of ills—perhaps even before the sufferer is born.

Genetic engineering of course implies more than counseling and therapy. Beyond negative and positive genetics, how about a superlative degree? If remedial work can be done, why not improve *beyond* the norm through genetic engineering? Shouldn't we engineer stronger, healthier, brainier human beings? Or perhaps we should content ourselves with implementing the age-old concept of eugenics, creation of a superrace through controlled breeding. This notion has intrigued men from Aristotle to Hitler to Hermann J. Muller.

Artificial insemination is three-quarters of a century old as an applied technique. Now artificial *inovulation* is helping obviate natural conception, and there is *ectoconception,* the fertilization of an egg in the test tube rather than the womb. Artificial wombs of glass and steel herald *ectogenesis* and the decanting of babies as in Huxley's Central London Hatchery. Control or selection of the sex of an embryo is now possible; before long parents may have whichever they choose. It even seems possible that a couple—or an individual—may have a child from sperm frozen decades ago, fused with an egg of the same vintage in a lab dish and belatedly implanted in a surrogate mother—human or stainless steel—for the nine-month period of gestation.

Perhaps the most remarkable prospect is the "Xeroxing" of human beings. Production of identical genetic copies of a donor (using the nucleus of one of his *mature* cells as a blueprint) by "cloning" has been achieved with lower animals, often in great batches of exact replicas. Such living carbon copies are as much like the parent as identical, one-egg twins are like each other. The ramifications of this conceptual tour de force are fascinating.

Among the "diseases" geneticists hope to cure with therapy or engineering is old age! Some experts on aging believe that growing old can be accounted for by a genetic clock whose gears or escapement we may be able to tinker with, toward a goal of Methuselah-like longevity. As a companion piece to this feat, other bioscientists are inching toward the artificial production of life, or as some prefer, the production of artificial life, which may be a unique form.

One of the most remarkable aspects of the genetic revolution is the quietness with which it proceeds. Despite such topical controversies as Women's Liberation, the IQ argument, the population explosion, and the question of race, the vital part genetics plays in each of these issues is largely not mentioned. Only sickle-cell screening has achieved much notice. Hardly a conspiracy of a formal nature, this soft pedaling of the subject of genetics is nevertheless serious enough to alarm James D. Watson, who with Francis Crick a few years ago tracked down the root of genetics in the DNA molecule of the gene. Watson and a handful of others have urged government to make the stuff of life off limits to the gene tinkerers. But he is realistic enough to admit that even such sanctions—which probably will not be forthcoming—would not stop the gene machine. Aboveboard or illicitly, some scientists will pursue the fascinating prospect of manipulating the mechanisms of life. It may turn out that only the genes themselves can draw the line beyond which their stewards cannot go.

GENETIC REVOLUTION

The human individual first comes into existence as a minute informational speck which resulted from the random combination of a number of still more minute informational specks derived from the genetic pool which his parents passed on to him at the moment of impregnation. His subsequent pre-natal and post-natal development may be described as the process of becoming what he already is from the moment that he was conceived.

Paul Ramsey, *Fabricated Man*

1   Tracking Down the Gene

Scientists who study the phenomenon of memory are searching diligently for the "engram," a tangible trace that will make it possible to come to grips with the workings of the mind. Although the term was coined by Canadian scientist D. O. Hebb decades ago, searchers have not yet isolated the engram: they are hampered because they do not really know what it is they are looking for. Indeed, according to Roger Sperry at Caltech, they wouldn't recognize an engram if they came across one in the flesh. To great extent, this was the problem of scientists who set out long ago to isolate and identify the "seeds of mankind," the hereditary entities that give us our reproductive immortality, our species similarities, and our individual differences. Finding a needle in a haystack can

1

be complicated by ignorance of what a needle is, or even that one exists.

A century ago the journal *Scientific American* plaintively described the problem:

It is thought by the advocates of the physical school that although nature can be built up by the play of the ordinary chemical forces, at some future time, when we shall know far more of molecular physica than we do at present, we may hope to explain how it is done. This is the cherished hope of modern evolutionists, and of the advocates of the physical theory of life. From the hugest plant and animal on the globe to the smallest organic speck visible under the microscope, all have been built up, molecule by molecule, and the problem is to explain this molecular process. Here in this region the doctrine of natural selection and the struggle for existence can afford no more light on the matter than the fortuitous concourse of atoms and the atomical philosophy of the ancients.

The search for the spark of life is as old as Democritus' atomist philosophy and the Promethean myth. Understandably, this has been the most elusive grail ever sought, and until recent times man had not even sighted it. Now, suddenly, he has his hands on the mechanism of life itself. The tracking down of the responsible entity, the DNA molecule that makes up the gene, is a fascinating tale that spans more than two thousand years of time and the efforts of scores of men, some of them clever hunters and others blunderers along the track. As in all science, today's achievers stand on the shoulders of those who went before.

The Ancients

The student of genetics generally is told that the science began precipitously in about 1900, with the "rediscovery" of Gregor Mendel's findings of heritability in plants. This gives the impression that genetics, the study of what makes us and changes us, dates back little more than seventy years. However, Mendel did his monumental work with sweet peas and other plants more than a century ago. A century before that the French encyclopedist Denis Diderot published a story called "The Dream of D'Alembert," in which society *grew* human beings in pots labeled "scientists," "artists," and other varieties. This was some 150 years before Aldous Huxley jolted readers with his *Brave New World* of decanted Alphas and Betas, humans mass-produced on order in a "hatchery" devoid of romance, love, or even sexual intercourse. And Diderot followed by 2,000 years a Greek speculator on human origins.

Obviously we need not know the latest findings of geneticists to wonder about our beginnings, why human beings beget human beings, why mice produce mice, and why a black and a white parent don't produce in each offspring a blend of their colors but sometimes black, white, and brown in the same family. Aristotle did his science several centuries before Christ, and among his studies were human origin and heredity. Many of his theories were wide of the mark; he was not aware that plants were male and female, for example. But he made some shrewd guesses on evolution nevertheless.

We tend to remember Aristotle principally for his philosophy and his system of logical reasoning, but he was also an eminent biologist, and the best of his 150 volumes treated this science. He classified 520 species of animals and actually dissected about 50 of these to add to his knowledge of living things. He pioneered the system of organizing species into hierarchies. As might be suspected, he also made some horrendous mistakes. He opposed the theory of Democritus that all matter was made up of tiny, indivisible particles called atoms, this being a step too big—or too small—for Aristotle's objective view. He could not see an atom, so he could not believe in one. (Neither could the brilliant physicist Ernst Mach, more than 2,000 years later.) Democritus, incidentally, proposed a "mosaic" theory about heredity in which a child could be male and yet inherit some of his mother's characteristics. Here was a crude foreshadowing of modern genetic theory.

Strangest of all was Aristotle's assessment of the brain as merely "an organ to cool the blood"! However, he believed that the world was round (when he walked north, new stars came into view), and he classified the dolphin as a mammal because it brought forth living young attached to a placenta. No other scientist sided with this biological heresy until more than 2,000 years later when German biologist Johannes Muller agreed with Aristotle's classification.

Aristotle reasoned that there was a chain of progressive change in living things (although he incorrectly believed that some creatures were created spontaneously), and thus was one of the first to entertain the notion of evolution from common origins. He envisioned a "ladder of nature," moving from inanimate matter through plants, lower animal life, egg layers, and mammals, to man. All life, he thought, was guided by divine intelligence.

Although Aristotle had studied the embryo, and knew that this was the developing creature, he could not visualize the egg that produced it, or the tiny sperm cell with its vital triggering contribution. Able to see the whole sweep—he knew the earth was round!—he couldn't imagine the tiny building blocks that made it up. A man who could not accept atoms of rock or metal would be hard put to conceive, much less discover, such entities as human reproductive cells.

Hippocrates, who died before Aristotle was born, advanced a "two-seed" theory of reproduction: semen mixed with vaginal secretion, and during pregnancy gradually coagulated to form the baby. This theory had a logical appeal, but Aristotle rejected it in favor of a "one-seed" mechanism of reproduction. He believed that semen from the father merely "organized" the blood in the womb, which became the baby. While there was a germ of truth in this notion, Aristotle was still far from the actual mechanics of reproduction. So was Aeschylus, whose theory held that the father was the actual parent, and that the mother was only "nurse to the life sown within her."

The Reproduction of Life

As ours may to future generations, the ignorance and naïveté of early man seems incredible today. Not many centuries ago, for example, the "scientist" Paracelsus could in seriousness prepare a recipe for creating human life in the laboratory: let some semen putrefy, nourish it with human blood in a proper temperature, and in short order you would have a homunculus, or tiny man. "Get with child a mandrake root," wrote John Donne. Although Donne likened getting one pregnant to the difficult feat of catching a falling star, many people of his time still believed that the mandrake was indeed a miniature living creature. It was also accepted as true that lambs grew out of the ground, and that a tree in Scotland overhanging a river dropped leaves into the water that became fish, while those that fell on earth became birds. It is often suggested humorously that the parents of large families don't really know what causes the children. Once this was generally the truth of the matter, and even now there are backward peoples who attribute pregnancy to spirits, nature, accident, or just about anything except the fertilization of an egg by a sperm.

A basic controversy developed early over "preformation" and "epigenesis." Preformationists believed that the seed of the male was in fact a duplicate man so tiny it was invisible to the eye but slowly grew until it resembled the father—and of course carried other seeds to repeat the process ad infinitum. Wiser heads developed the counternotion of epigenesis, the creation of the embryo from raw materials in the egg. The battle alternately raged and languished over the centuries. Preformation expanded to "encapsulation," the Chinese box theory that the first egg contained in miniature all those who would subsequently be born, nested one inside the other and each proportionately smaller.

Leonardo da Vinci made accurate drawings of the act of intercourse, although his mysticism produced some strange errors including an extra urethra that pumped "soul" from the father's spine. Da Vinci and some others knew the mechanics of conception, of course. But dissectionists familiar with most of the body's parts were hampered by their own false assumptions as well as the ignorance, prejudices, and dogma of the times. Da Vinci tactfully concealed many of his drawings and speculations so that they were not available until quite recently.

Andreas Vesalius was the "master of anatomy," and his dissections revealed practically every detail of the body: bone, muscle, blood vessels, and nerves. He left plates that even today are marvels of accuracy and detail. Here were miracles of the human body in plenty, but the greatest secrets remained invisible to the naked eye and resisted all attempts at scientific guesses. Even the circulation of the blood was still a mystery up to the time of William Harvey, and scientists were almost as far as ever from understanding reproduction.

The problem of finding the secret of life was not merely the ignorance of the searchers. A real and forceful barrier was the opposition of the world generally to any prying into such vital secrets. Orthodox science, the Church, and a superstitious citizenry all were potential enemies of any new discoveries that would challenge old beliefs. Harvey withheld his theory of the circulation of the blood for thirteen years, fearing reprisals from the scientific community, the Church, and his own patients.

Yet it was Harvey who published a paper advancing the theory *ex ovo omnia,* "All from the egg." Today most people are aware that mammals, like birds, develop from eggs fertilized by sperm. Harvey proved by dissecting pregnant does that Hippocrates' coagulation theory was false, but left himself with the mystery of fertilization. Although the human egg had not been discovered, Harvey was sure it must exist. But what accomplished the actual fertilizing, if semen was not responsible? The sperm itself had not yet been discovered, or even thought of, and Harvey died "at a stand" about the actual process of generation.

Part of the problem was the egg itself. Even Hippocrates had investigated the developing embryo in a chicken's egg, and this procedure was repeated by experimenters down through the centuries to and beyond the time of William Harvey. A chicken egg is sizable, and thus comprehensible. A human egg, on the other hand, is all but invisible; who could believe that such a mite holds the future of humanity? Progress came inchingly with dissection of animals, like Harvey's does and the dogs used by other investigators. It was even whispered that Cleopatra had ordered similar experiments performed on pregnant slave girls.

The great French scientist Georges Buffon muddied the waters with a colossal mistake in laboratory analysis. He stated that a dissected unmated bitch was found to contain particles identical with those in the semen of a male dog, and thus the female alone was responsible for reproduction. Buffon's explanation was to quote a medieval theory about the germs of life "dropping from the stars." The air about us, he said, was filled with minute living spores that were taken up by animals and produced sperm. Years later, Swedish scientist Svante Arrhenius would suggest his "panspermia" idea in all seriousness, claiming that life spores traveled across the vast emptiness of space under pressure of the light from stars.

Although much of the modern theory of genetics had been at least hinted at, most people in the eighteenth century still believed in "spontaneous generation"—that living creatures just grew from dung, flour, grain, and even old rags. In 1668 Francesco Redi had done experiments that led him to believe that all living things must come from seeds of the plants or animals themselves, rather than arise spontaneously from the air or matter. But few listened to Redi. Lazzaro Spallanzani had to repeat these experiments to get the attention of scientists. (Incidentally, Spallanzani's experiments of boiling meat to kill bacteria led a French chef, François Appert, to invent the process of canning soon after 1800.) Another Frenchman, the great Louis Pasteur, later added his prestige to experiments proving that spontaneous generation was not taking place on earth—at least not in those places where he prepared his flasks. It seemed proved that life had to proceed from life, rather than arise from inert matter.

There is a great riddle still asked about the priority of the chicken or the egg. English novelist Samuel Butler, who had his own mystical theories, said the question was irrelevant but that the hen was only the egg's way of creating more eggs. In other words, the important things to nature are the sex cells. Nevertheless, curious medical men found the egg, generally that of the hen, a rewarding study ground for the developing embryo. Incubated eggs could be broken open at regular intervals to follow the chronological development of the chick inside. In 1836 an embryologist published a book in which he described his attempts to "create" a chicken by treating eggs with various fluids including vinegar and oil, and also mixing rennet with egg white. However, such test-tube fertilization failed to produce anything more than omelets.

We noted that while Harvey believed all life came from the egg, he died not sure what else was involved in the process. Neither was anyone else until 1875 when German scientist Oskar Hertwig watched through a microscope and saw sea urchin sperm migrate to an egg of that species and fertilize the egg. Now it was clear that there *were* two seeds involved in the production of living mammals: a female egg and a male sperm. But just as the "ultimate particles" of Democritus were only the gateway to an incredibly tinier world, so were those microscopic living cells that produced animals and men. An even knottier puzzle now presented itself, and new tools would be needed to unravel this one.

The Living Cell

Einstein solved the question of relativity without laboratory experiments, relying on thought alone. Perhaps so keen a mind might also decipher the mysteries of evolution, generation, and invariance in living things. But in fact it was only with the microscope as an aid to seeing the living mechanisms involved that men of science could begin to tackle the great problems of life that tantalized the curious for centuries.

Galileo peered through his telescopes into the heavens, investigating the huge universe about us. He also turned lenses inward on the microworld, and after 1600 the microscope was a well-known instrument. But a linen draper in Holland named Anton van Leeuwenhoek would build better microscopes and with them see things no one had yet seen—or would see again for more than a century.

In 1665 Robert Hooke in England published a marvelous book called *Micrographia,* filled with sixty plates done from his observations under microscopes made of droplets of fused glass. It was Hooke who coined the word "cell" for the tiny compartments he observed in cork. They looked for all the world like monastery cubicles, and they were indeed the dead walls of once-living cells. Leeuwenhoek also looked at cells under his ground glass lenses, but he saw far more than empty walls. The Dutchman identified red blood cells, and was the first to see single-celled protozoa, the amoeba common to science classes today. He wrote of taking a "bit of white matter" from his teeth and observing in it "tiny animalcules, very prettily a-moving." Amazingly, some of his animalcules must have been bacteria, hitherto unimagined entities seen for the first time by a tradesman who was not a scientist at all in the formal sense.

Leeuwenhoek also correctly identified human sperm. A medical student who first saw these in the semen of a patient with venereal disease dismissed the tiny, threadlike cells as a product of the infection. Leeuwenhoek with his natural curiosity and keen logic correctly assigned to the sperm its true nature. There were, he announced with awe, more sperm cells in the tiny sample he used than there were humans on earth! Soon he had found analogous sperms in the milt of fish, and the semen of horses and other animals.

Many scientists still believed that tiny humans resided in the sperm, or in the egg, that preformation was the mechanism of generation. Some actually claimed to have seen mite-sized donkeys, roosters, and horses in the sperm of those animals. Leeuwenhoek, although believing the preformation theory, reported he was not able to see these homunculi; and a wag named Dalenpetius ridiculed preformationists by claiming he had seen one homunculus take off its little coat under his microscope.

Vesalius had done masterful dissections of the human body. Jan Swammerdam, with the microscope, used his scalpel on infinitely smaller life forms in the seventeenth century. For example, he dissected the nervous system of the honeybee and also the mayfly larva, a tiny thing. Exacting as these feats were, they were gross operations compared to probing individual living cells.

Robert Hooke had published drawings of plant cells in 1665, but the idea and the importance of cells in animals and plants long escaped science. Nevertheless, a remarkable observation was made by German naturalist Lorenz Oken in 1805: "All organic beings originate from and consist of vesicles or cells. These, when detached, and regarded in their original process of production are the infusorial mass or primeval slime whence all larger organisms fashion themselves or are evolved." Here, in two sentences, was cell theory, with evolution thrown in for good measure! But it would be decades before anyone scientifically pursued the idea.

In 1830 botanist Robert Brown discovered a dark blotch in cells he studied, but it was not appreciated that this was the "control center" of the cell, or even that one was indicated. Not until 1839 was cell theory formally proposed. The German botanist Matthias Jakob Schleiden had reasoned that plants were made up of cells, each of which had an independent life and could be separated from the others. As Schleiden described this to another scientist, Theodor Schwann, the latter suddenly realized that animals too consisted of countless tiny cells. In 1838 Schwann published his paper "On the Correspondence in the Structure and Growth of Plants and Animals."

That cells were the building blocks of life was certain to biologists by the middle of the nineteenth century. Here seemed miracle enough, but there was far more to life than that. Embryologists had long watched with awe the rapid development of a baby chick. Almost before the eyes of an observer, a blotch of red became a functioning heart. So too with the other organs and parts. Now it was known that this was a gross simplification of what was actually happening. Even a tiny chick's heart was made of countless cells, and each of these seemed to have discrete parts within them to make the cell itself huge by comparison. Something must be directing the growth of cells, which in turn led to the development of the plant or animal. But man could not understand the driving force and control of the cell.

A century after Schleiden and Schwann's cell theory, Dr. Alexis Carrel would say with reverent wonder in his book, *Man, the Unknown:*

An organ builds itself by techniques very foreign to the human mind. It is not made of extraneous material, like a house. Neither is it a cellular construction, a mere assemblage of cells. It is, of course, composed of cells as a house is of bricks. But it is born from a cell, as if the house originated from one brick, a magic brick that would set about manufacturing other bricks. Those bricks, without waiting for the architect's drawings or the coming of the bricklayers, would assemble themselves and form the walls. They would also metamorphose themselves into windowpanes, roofing-slates, coal for heating, and water for the kitchen and the bathroom. An organ develops by means such as those attributed to fairies in the tales told to children in bygone times. It is engendered by cells which, to all appearances, have a knowledge of the future edifice, and synthetize from substances contained in blood plasma the building material and even the workers.

While the origin and other secrets of life were sought under microscopes and in the dissecting laboratory, some scientists approached the problem in a different way. Aided by new knowledge about geology and the age of the earth, Buffon and Lamarck in France proposed theories of evolution. Lamarck's "acquired characteristics" fought a losing battle with the newer "Darwinism" that saw natural selection as the mechanism of evolution. Darwin, however, was unsure enough of his theory to later include a naïve "pangenesis" hypothesis to accommodate the supposed acquisition of characteristics from the environment. As a compromise, Darwin conceived hereditary entities in the bloodstream called "gemmules," which altered the reproductive cells so they could pass on acquired characteristics.

Despite the great impact this new theory of natural selection had on science, society, and economics, grand theorizing or overviews were of little help in explaining on the biological level how life was reproduced and how life forms were altered, as *Scientific American* lamented in 1872. It was necessary to get down to life itself for practical applications of heredity. No one had yet cut open a reproductive cell and analyzed its functions; plant and animal geneticists had to work backward, treating the mechanism as a black box whose properties they inferred from the progeny of known parents. For there was already an accurate mathematical study of plants, a study that would point the way to the factors involved in heredity.

Mendel's Factors

In 1865 the first "geneticist" read a paper detailing the results of thousands of experiments that proved beyond doubt that there was in cells a detailed blueprint that determined "the future edifice," as Alexis Carrel later called it. This pioneer bioscientist was Gregor Mendel, an Austrian monk who had carefully tended many generations of plants and cataloged the results of the crossing of various types.

Heredity, the passing on of characteristics from parents, was not a "fluid blending" process as had been thought. It was not like a mixing of red and white paint to get pink, or of black and white to produce gray. For one could mix red and white flowers and get all reds or all whites, or some of both, and some mixtures of colors. The hereditary "factors," as Mendel called them for lack of a better word, did not fuse to produce changes. They were more like marbles that retained their individual identity and could pair up in all sorts of combinations. Because of this discrete nature of factors, the laws of statistical probability could make remarkably accurate predictions of the outcome of various crossings, including, it would be learned, those of human beings. But not for some time was this potential exploited.

It might be said that Mendel gets too much credit, that he was merely the one lucky enough to put all the pieces together and complete the solution of an ancient puzzle. Certainly plant and animal hybridists long before him had conducted experiments and kept records of similar nature. Mesopotamian clay tablets thousands of years old document what historian Lancelot Hogben has called "the first Stud Book in history." This chart of pedigreed horses includes the symbol for female that is still used. Thomas Knight, president of the British Horticultural Society, was among those who realized how much more quickly hybrid results could be produced with plants than with animals. An experimenter with animals, he also crossed fruit trees and flowers (including sweet peas) during the period from 1787 to 1823. Mendel was born in 1822.

Perhaps Mendel was a Johnny-come-lately in plant genetics. Certainly he reaped no rich rewards during his own lifetime for his monumental work. Mendel made about as much impression on the waiting world as Oken, who had earlier hinted at evolution. His audience yawned and went home. Eminent scientists working in the same field dismissed his results as having no significance. It was not until thirty-five years later, long after Mendel, and Darwin too, had died, that three other men reached similar conclu-

sions about invisible "heredity factors." Belatedly they found that
Mendel had long preceded them, to gather nothing but dust on his
papers in the library at Brünn, Austria.

Life's Blueprint at Last

German biologist Walter Flemming in 1880 learned he could
dye that vague splotch detected in the cell by Robert Brown fifty
years earlier. The dyeing process made visible some interesting
things: at the moment of cell division, tiny threadlike bodies
formed in pairs in the cell nucleus. The Greek word for color was
used to describe the process, and the threads acquired the name
"chromosomes," or colored bodies.

Chromosomes are visible under the microscope. It was learned
that humans had 48 of these, arranged in pairs—or that was the
assumption until 1953, when the number was corrected to only 46
(unless there is a serious hereditary defect in a person). But 46
"factors" seemed a handful to cause all the things that happen in
the growth of an individual. Researchers began to infer, without
seeing it, the "gene" or blueprint of life. American geneticists including Thomas Hunt Morgan and Hermann J. Muller began
studying fruit flies (which have large chromosomes) and learning
that the genes are tiny components of chromosomes.

In 1869 German chemist Friedrich Miescher isolated an unknown substance in cell nuclei and gave it the name nucleic acid.
Sixty years later the American biochemist W. M. Stanley isolated a
strain of tobacco mosaic virus made up not of cells but chromo-
some-like fragments. Containing Miescher's discovery, nucleic
acid, this virus could make a copy of itself inside the cell it invaded. With the tobacco mosaic virus, science had almost isolated
the gene, the blueprint of life. In the 1940s the DNA revolution
came when it was learned that nucleic acid carried information
that controlled and directed cell growth. By the early 1950s the
guesses of Linus Pauling and work by Maurice Wilkins, James

D. Watson, and Francis Crick led to identification and mapping of the gene itself: the miraculous "double helix" of deoxyribonucleic acid, the giant DNA molecule.

In the resulting genetic revolution, man has further zeroed in on these hereditary particles which decree that human beings give rise to human babies, and fish to little fish: the genes which also make possible the slow, evolutionary change in living things.

Tracking down the gene has taken more than 2,000 years, then, and the lifetimes of many curious and dedicated men, willing to sacrifice their time, their careers, and their substance for science. It has taken the development of the optical microscope, and more recently the electron microscope and sophisticated techniques like X-ray diffraction, to isolate the invisible gene. Even a single gene is an awesome assemblage of chemical and physical components; a human cell has thousands of them. And each gene includes a chemical alphabet that can code a fantastic number of instructions. It has been estimated that there are more possible combinations of human genes than there are electrons and protons of physical matter in all the universe.

As human beings we tend to overcomplicate problems. Life is a basically simple process. Animals, with no knowledge or concern about the process of life, automatically go on being a part of it. But human beings are curious, with a curiosity that kills them at times. For geneticists, and many others, the riddle of life is the most intriguing riddle of all. Some believe that God put it all into motion, and continues to drive the system. Others see the universe as a Godless natural system of physical entities that have blindly evolved into the most complex living things. In either case, however, it is the gene that foretells, builds, and operates the engines of life. The gene, as far as we know, is the ultimate particle of life.

> Genetics: The science that is concerned with biological inheritance—that is, with the causes of the resemblances and differences among related individuals.
>
> McGraw-Hill *Encyclopedia of Science and Technology*

2   Genetics: The Biology of Development

In all the thousands of years man searched for the seed, the master pattern of life, that seed was functioning just as it functions today. Genetic control gives variety to the world, yet it gives order too. The life force running amuck as it does in cancer hints at the chaos of a world without proper genetic control. For the gene is not only the law but the enforcer of law as well. Living creatures are physicochemical computers, if you will, and the gene is both program and control center for what amounts to a chemical factory.

According to evolution theory, the process of bisexual reproduction and its consequent uniquely different creatures evolved by blind chance. The first life forms reproduced by simple cell fission, with "daughter cells" sharing hereditary endowment from the original cell. There was no mixing, no blending, no change all down the line except what might be produced by mutations. In bisexual reproduction, however, the mixing of genes can result in great differences among progeny. The implications of the "gene pool" are tremendous, accounting for great variations in the human race since man first appeared on earth.

We come from exceedingly small beginnings. Geneticist Hermann J. Muller once estimated that *all* the eggs from which the human population had sprung (at that time 2.5 billion) would occupy a volume of less than one gallon. And the sperm cells which fertilized them would fill only as much space as half an aspirin tablet! Theodosius Dobzhansky has estimated that all humans alive today were genetically structured with only about 20 milligrams of DNA. On this scale, a single ounce of DNA would suffice for about 4,000 billion people! But there is much more to it than that. Since the first humans, an estimated 100 billion of us have walked the earth, and most have been recognizable copies of the first man.

Every one of those hundred billion stemmed from the *original* human genes, subject to whatever mutations or changes have occurred in the millennia that have elapsed. Thus the hereditary destinies of all those billions of people were spelled out in the genetic material in the sex cells of the first human beings.

If the beliefs of evolutionists are correct, even this miracle is surpassed. If man is but the ultimate flowering of a "ladder of life" that began with the protozoan, then the first genetic material spawned all the uncountable forms of life that have lived since then and will live in the future. There are now about 3.5 billion human beings on earth, but man is only one of some 1.5 million species known so far (and new ones are being discovered at the rate of 10,000 more a year). Here is a miracle of reproduction to make the old theories about nested Chinese boxes simple by comparison.

Reproduction and Development

Fire has been important to human beings for tens of thousands of years. There was a time when man did not know how to create fire anew but had to keep alive an ember taken from fire caused by lightning or some other act. Man eventually learned to make fire "from scratch," but despite Sunday-supplement writers, we still cannot create life anew. We can only reproduce existing life that resides within us.

At the time of conception, a male parent contributes a single cell—the sperm—which merges with the female's egg cell. Each of these cells is already living, of course, but each on its own would survive only a short time if union did not take place (or the cells were preserved by freezing or some other method). A human sperm cell has a head about two microns in diameter (a micron is 1/25,000 of an inch) and four microns long, plus a tail 50 microns long. A human egg cell is about 1/100 of an inch in diameter; we would be hard put to see one without a microscope. Yet the egg is about 85,000 times as large as the sperm!

Once joined, these tiny entities set in motion a chain of events that leads ultimately to the production of some 50 trillion cells, in some cases weighing 200 pounds or more. Such a phenomenon might be compared with the chain reaction of nuclear fission, but this is a rough analogy at best. Trigger a nuclear blast, and stupendous energy is released. But human nuclear fission requires far more sophisticated guidance than the most complex nuclear explosion. And this guidance must come from *within* the nucleus of the cell.

After the original cell merger initiates new life, the cells divide through normal binary fission; the nucleus splits its chromosomes lengthwise and produces identical nuclei for each cell. The growth of life from the original cell is an awesome miracle. Books have been written detailing this majestic process, and those who have not read such descriptions should do so. We are concerned here with the operation of the genetic blueprint in directing the creation of new cells, and their subsequent growth and specialization into their destiny.

Here is Carrel's "magic brick" in action, a brick that contains the potential for the entire building. A cell is a tiny living thing in its own right, equipped with miniature "organelles" analogous to the larger organs in living creatures. Genes direct the machinery of a new cell so that it forms these control organs, surrounds itself with a membrane, grows to just the right size, takes in nourishment from the environment, throws off wastes, and at the proper time splits itself into more cells. Genes also inform cells whether they are to be skin cells or heart cells, how many of them there should be, their correct configuration, and so on. Genetic control starts the various parts of the human body at the right time, grows them to the proper size and shape, then makes them stop growing. Genes program the cells to replace themselves as some wear out, and at the proper time to produce sex cells for a new cycle of life.

Of great importance is the feedback system which reacts to the outside environment. Forty-six human chromosomes shape a creature flexible enough to live on the Arizona desert, the Himalaya Mountains, and—with some technological assistance—even on the moon. Human beings can withstand heat or cold, high pressure or low, a dry environment or a wet one. Simpler creatures tolerate a far greater environmental range. Evolutionary theory denies the existence of acquired characteristics, the working of the environment directly on life to change it to better match the environment. Thus the genetic mechanism of the *original* living thing must have encompassed all this capability.

People in developed countries can look forward to a life of about seventy years, barring accident or disease. All of this is a time of development, and all of that development—all the millions of heart-beats and lung movements, all the coursing of blood through veins, all the cycles of digestion, and all the mental activity—is under genetic direction, the direction that told the egg how to divide and grow from one to 50 trillion, and when to stop. Some geneticists believe that when we grow old and die, it is only because we have run out of "genetic program." In a later chapter we will look at the problem of aging and its genetic implications in more detail.

Molecules and Cells

A rock or chunk of iron is an inorganic substance. So is water. Inorganic molecules are relatively simple in structure, consisting of only a few atoms. Organic molecules are more complex. There is an understandable misconception about the term "organic," and many believe it means "containing life." Organic compounds need only contain carbon to qualify, however. Sugar, consisting of carbon, hydrogen, and oxygen, is organic. So is rubber. Thus, not all organic molecules are living. But all living molecules are organic.

There are believed to be ninety-two natural elements; scientists have created enough artificial ones to increase the total to more than a hundred. Twenty-four of the natural elements are found in living matter; perhaps there are others not yet detected. But life seems to require only six basic elements: carbon, hydrogen, oxygen, nitrogen, phosphorus, and sulfur. From these few are made up the complex giant molecules called proteins. Nutritionists stress the importance of protein in our diet, and the wisdom of this advice is evident in the fact that protein is the stuff involved in building and maintaining a human—or any living creature.

When biologists began to "tease" individual cells out of living tissue they may have thought they were probing bottom on the scale of life. But within the cell lives the even tinier nucleus. That nucleus in turn is huge compared with the threadlike chromosomes it holds.

Chromosomes

In most of our trillions of cells there are 46 chromosomes, all of them responsible for various physiological characteristics. In sexual reproduction male and female cell nuclei fuse into one nucleus. This would seem to result in 92 chromosomes, and the next union would produce 184, and so on. Such an abundance of genetic material would lead to problems, so nature has curbed it by the process of meiosis, in which each sex cell at union contributes only 23 chromosomes. Their joining to produce 46 is not a fluid mixing, but a combination that may produce a child like the father, like the mother, like both, or entirely different.

Nearly all living creatures have chromosomes. Man's 46 are toward the upper end of the numerical scale, where pride would assign them. However, some monkeys have 54 chromosomes, and cattle have 60! Even the lowly rat has 42. Onions have chromosomes too, 16 in all; lilies have 24. The fruit fly—which geneticist Thomas Hunt Morgan said God must have made for Morgan alone —has only 8, making genetic experiments easier. Some insects have as few as 4, but the marine protozoa *Radiolaria* boasts more than 800 chromosomes. Apparently it is not quantity but quality that counts.

Chromosomes pair up. One pair determines a most important characteristic of humans, that of sex. In a female both sex chromosomes are X's. In a male there is one X and one Y. Thus while women are "all female," men seem a blend of both sexes. Intuition suggests that if women carry XX sex chromosomes, men should carry YY. However, some thought will indicate that this arrangement would lead nowhere. There would be no way to produce any kind of offspring but a hybrid with one of each type of chromosome.

Forty-six control agents could hardly govern all our physiological attributes; each chromosome must do much more than one task.

In the case of the sex chromosomes, for example, they also govern such things as baldness, a "sex-linked" characteristic, since baldness strikes far more men than women. Hemophilia, a condition in which blood fails to clot, is also a sex-linked genetic trait, affecting only men, although the disposition is transmitted through women, of course.

Working with fruit flies and their relatively simple genetic structures ("relatively simple" is an exaggeration best appreciated by Morgan, Muller, and others who devoted their careers to studies of these insects), geneticists have mapped the chromosomes and genes to show which of them control which characteristics. Other researchers are beginning to map similar links in people, a far more complex task. For example, in crossing certain insects and predicting the effects of such matings on progeny, C. B. Bridges (a student of Morgan's) found some offspring that did not breed true. He reasoned that the cause must lie in the possession by some female flies of three X chromosomes rather than the normal two. Investigation disclosed that such "superfemales" did exist, although rarely and precariously. Such anomalies occur in human beings as well, and we shall discuss this later.

In the words of geneticist Theodosius Dobzhansky, "An experienced geneticist can 'play' with the genes and chromosomes of *Drosophila* flies almost like a chess player with his chess figures. . . ." This understates the complexity of genetics, of course. A chromosome is a near-invisible filament, seen only with sophisticated microscope techniques. Yet this tiny chromosome is as complex as the Capitol in Washington. The Capitol symbolizes national government, but it is the thousands of individuals *inside* the government who are the real agents. Similarly, the chromosomes are actually assemblages of thousands (or tens of thousands) of genes.

The Gene Itself

DNA is a familiar acronym; nearly everyone has heard of it and has at least a rough idea of "the giant molecule of the gene," or the "blueprint of life." The gene's *deoxyribonucleic acid* is the Cinderella compound that confers prestige in the life sciences, and produces the most Nobel Prizes for biologists, geneticists, biophysicists, and other specialists in the new genetic revolution. "Giant molecule" suggests a sizable structure. It is, when compared with simpler molecules such as those of water or iron.

No one has seen a human gene; however, a decade ago Linus Pauling guessed brilliantly that the gene was actually a giant organic molecule in the shape of a helix, or corkscrew. Proof that Pauling was right came only after some clever detective work on the part of scientists in many laboratories, using several different approaches. The human gene is too tiny for even the electron microscope to make it visible. But X-rays have been sent through and around it, and by analyzing their deflection the structure of the gene was established—a technique like throwing rocks at a target hidden in a cloud of smoke and establishing its shape by the pattern the rocks make in a mud wall on the other side.

The gene proved to be not one helix, but two—a pair of corkscrews nested one within another, linked by tiny assemblies of amino acids that tie them together like the rungs of a twisted ladder. These rungs are "hydrogen bonds." Altogether four different amino acids (of a total of 22 that are known in living things) appear in the ladder rungs joining the DNA double helix. These are adenine and thymine, which form complementary links; and cytosine and guanine, which also link. Along one strand of DNA are protruding amino acids of these four kinds in what seems random order. Across from each is a complementary amino. Linkage of matching pairs of helixes is accomplished both by the physical structure of the amino acids and by the weak electrical attraction of the hydrogen bond.

The appropriateness of the double-helix structure can be appreciated if we recall the process of cell division and the necessity of passing on an exact genetic blueprint to the new cell. At the moment of division the double helix "unzips" into two strands, each retaining the halves of its ruptured hydrogen bonds. From "raw material" amino acids swimming about in the cell, each strand picks up corresponding bonds and makes a new matching helix partner. At least this is the present hypothesis, and it seems to meet the requirements of gene function admirably.

We have noted that bioscientists first studied genetic mapping with the chromosomes of fruit flies. They added to their knowledge by studying the simpler bread mold. Later they progressed down the scale of simplicity even to viruses. It was in these latter studies that nucleic acid was at last identified as the transmitter of genetic information, since viral DNA could duplicate itself inside a host organism, using the raw materials available in the cells.

The ultimate in genetic simplicity was found in the bacterial virus X174, with only a *single* strand of DNA. This *bacteriophage,* or "bacteria eater," does not replicate a matching strand of DNA until it enters the host, thus minimizing the load it must carry in its travels. Scientists had already come across this elegant simplification in the sperm itself, composed almost entirely of nucleic acid. The sperm is "half a blueprint" for life, and little more, able to exist only for the time it takes to reach the egg.

A bacterial virus invading a cell injects into it a "ribbon" of DNA analogous to the gene found in human cells; this genetic program or set of instructions then assembles copies of the virus from the protein "raw materials" in the host. It has been found with the electron microscope that this viral "program tape" can be 150 times as long as the virus itself. Estimates are that a single human cell nucleus may contain as much as five lineal feet of coded chemical instructions, a program tape equivalent to a thousand volumes as wordy as those of the *Encyclopaedia Britannica.*

Genetics abounds in homely analogies, for which we can be grateful. One of the best is that of the "master blueprint" which the plant foreman keeps locked in the safe at construction headquarters. His reason is valid: some worker on the assembly line might borrow the master drawing and accidentally set fire to it with his cigarette during a coffee break. So the foreman runs off copies of the blueprints, or pertinent sheets thereof, for his workers. Production is as good as ever, since the copies are exact duplicates.

Such analogies are often helpful in inferring what really goes on. It has been pointed out that they may also create errors in conceptual thinking and delay the truth because of their attractive simplicity. Conditioned to think that genes replicate like zippers, for example, a researcher might not believe his eyes when he saw one performing the feat more like a pair of hoops rotating at right angles to each other. Analogy is a tool only and does not necessarily describe the actual process; there are still many unsolved riddles in genetics.

At any rate, from the master DNA pattern in the nucleus, "template RNA" is produced, and it is this template or pattern that goes out and does the actual work of building cells, protein by protein. (RNA *is* slightly different from DNA, in that one of its four amino acids is uracil rather than thymine. At the moment, no one knows why this difference.) It is thought that the "pattern" RNA is inserted into cell organs called *ribosomes,* and that inside them it produces the proteins called for by the genetic code, in the exact sequence specified.

Nature's Code

Earlier we spoke of the need for a "program" to carry out the building of cells and their subsequent assembly in the proper order and numbers. For program, the geneticist says "genetic code." A code implies some sort of alphabet. That of the English language consists of 26 letters. Proteins are made up of 20 amino acids of various kinds. Surprisingly, however, it was learned that the gene contains only cytosine, adenine, thymine, and guanine. C, A, T, and G are the four letters in the DNA genetic code—or so it is presently thought.

An alphabet of 20 amino acids would yield 2 billion times as many "words" as an alphabet of only four aminos; why is nucleic acid and not protein used in the blueprint needed to make life? One suggestion is that because protein is a much more complex structure than nucleic acid it would be far more subject to errors in coding than the simpler alphabet. After all, the electronic computer uses only *two* symbols, 1 and 0, yet this simple code enables the Internal Revenue Service to process the tax returns of tens of millions of Americans, with different names and circumstances, and varying sums of money involved.

So far we have used the analogy of an alphabet forming "words" to direct the life processes of an individual. In real life, there are no letters or alphabets or words as we are familiar with them, but chemical words. It is enzymes that amino acids create, and one enzyme produces one protein.

Working out possible codes for the gene was a long and complicated business, full of false leads and disappointments. Geneticists have still not reduced the problem to direct translation, and may never be able to do so. However, in the decoding process some interesting possibilities were turned up concerning the mechanism of genetic mutations. The triplet code word or "codon" for glutamic acid is UAG: adenine, guanine, and uracil on the RNA chain in that sequence. The code for valine is UUG. If a uracil base is somehow substituted for adenine, the gene produces a different enzyme and protein. Such an error mechanism is thought to be involved in diseases such as sickle-cell anemia.

It is believed that the human body makes use of at least 10,000 different proteins, and perhaps as many as 100,000. This seems a tall order even for the miraculous genetic factory, but an easily conceivable molecule of amino acids could produce 20^{100} protein-specifying enzymes, a number beyond the comprehension of most of us. And if we are concerned that the genetic factory cannot operate fast enough, consider that in the laboratory mere men have succeeded in synthesizing RNA at the rate of 500 million copies in 15 minutes!

Considering only the chromosomes, there are fantastic numbers of different combinations that each individual may inherit. But it is the genes, or at least blocks of genes, that do the specifying. With only 250 gene differences among people, the number of theoretically possible gene combinations, and thus unique individuals, is of the same order as that of the electrons and protons in the universe.

The New Genetics

Gregor Mendel's "factors," then, turn out to be tens or hundreds of thousands of genes, each of which can produce an enzyme and in turn a needed protein. We already know of some 1,000 enzymes, and more are being found. Perhaps not all the genes are used; some may be duplicates or standbys. Some may be awaiting their need in the future. There are suggestions too that there are basically different types of genes. For instance "structural" genes are needed for the actual building of cells, but these can't work automatically. Thus the "operon theory" calls for "regulator" genes that use inducer and suppressor substances (probably histones and hormones) to start and stop the operation of structural genes.

Even before the discovery of nucleic acid and genes, early geneticists were able by experiment to infer much of the action of genes. Genes might work separately or with other genes to produce an effect. For example, possession of a certain gene might result in a flower being red. But if there were also another gene of a certain type in the chromosome the color would be but pink. Such joint-acting genes are known as alleles.

Genes also have the property of being "dominant" or "recessive." Genes of equal value may produce blending effects, such as a pink flower from genes for red and white. But some genes override the effects of others. We know, for example, that a parent with eyes of a certain color will pass on genes that result in the child's having that eye color regardless of what genetic material the other parent contributes. Recessive genes are those which singly produce no effect. When both parents pass on such genes, however, the recessives become dominant, and in some cases lethally so.

Mendelian genetics was a milestone in learning about life, but the new "molecular genetics" gave seven-league boots to the molecular biologist. This new discipline, cynically called "the practice of biochemistry without a license," is beginning to strip away the vagueness and mystery from the black box of genetic theory. No longer must the biologist work backward, painstakingly inferring cause from effect without ever observing the engine that makes it all happen. The geneticist has shed his clumsy gloves and now wields a microscalpel. For statistical probability he has substituted the physical and chemical nuts and bolts involved in heredity. At the same time, however, for difficult mathematical analysis he seems to have substituted a near-impossible physicochemical riddle.

For example, the T4 virus is only about 1/250,000 of an inch long, yet its genetic material includes about 100 genes of which about 20 have been located and identified. Only four have had fine structure analysis done on them. Human genes are estimated at between 50,000 and one million, of which only about 100 have been identified. A few have been crudely mapped, none has been analyzed as to its fine structure.

Chance and Error

Among the miracles of the genetic mechanism is its reproducibility. While there are mistakes now and then, and some abnormal human beings are born, most are normal. It has been suggested that a Cro-Magnon man dressed in a business suit and carrying a briefcase could mingle unnoticed with Madison Avenue crowds. This interesting contention may or may not be true. However, we are talking about tens of thousands of years of human history, during which tens of billions of humans have been produced from the original genetic blueprint. While all these individuals have been unique (even identical twins), all have been and still are distinctively human. Humans still produce little humans, elephants produce baby elephants, and mice baby mice. How is the genetic code so well guarded?

Redundancy, or surplus information in the genetic code, may be a safeguard mechanism. There also seems to be a "fail-safe" concept embodied in the genetic blueprint. It is very difficult to get at the gene to alter it; when this *is* accomplished the result is more likely to be catastrophic than subtle. As Darwin was aware, most mutations are not beneficial but harmful and often fatal. Miscarriages may terminate abnormal pregnancies. Thus when serious genetic error occurs, it is seldom passed on to descendants. The Thalidomide tragedy is an example. Violent chromosomal damage was done, producing children who if they survived were grossly malformed. Many will not mate to pass on genetic defects. For those who do, the bisexual genetic system may eliminate or screen the defects. Not all bald-headed men produce children with a like tendency. Not all mental defectives produce retarded children (neither do geniuses always produce geniuses).

It has long been held that radiation is a cause of mutations; that cosmic rays, for example, are to blame for freak births. Man himself has been adding radiation to the natural environment in which human life has evolved, and many scientists fear that nuclear fallout has long been affecting the genetic structure of people all over the earth. Linus Pauling, who suggested the structure of the gene, has said that perhaps as many as 140,000 deaths and as many harmful mutations have occurred as a result of this radiation. There are other man-made mutagenic agents as well, including chemicals.

If genetics underlies the conception and development of individuals in a species, it follows that in genetics we may also find the mechanism of development on a vastly broader scale—that of the origin and evolution of life. In the next chapter we shall consider genetics from this viewpoint.

The lines are fallen unto me in pleasant places;
yea, I have a goodly heritage.

Psalm 16:6

3   Origin and Change

Theories of evolution all imply evolution *from* something, a begin-
ing or origin. The search for the gene moved *forward* in time, over
a brief span of some hundreds of years. The riddle of the origin of
life is another matter, extending backward in time and hampered
by a dearth of factual evidence. It was difficult enough locating the
gene, with it right under our noses. But the beginning of life, however it came, was under no one's nose.

Later we will discuss the possibility that life migrated here from
some distant planet; to begin with we simplify by considering only
the appearance of life on this globe that is our home. Thus we can
at the outset put bounds on the age of life: it can obviously be no

older than the earth itself, and if we can determine how old the earth is, we can establish a maximum age for life as well.

The Origin of Life

Until about a century ago, it was commonly thought that the earth was only a few thousand years old. This belief stemmed from a strict chronological interpretation of the Bible; Bishop Ussher carefully added ages all the way back to Adam and established the date of creation as 4004 B.C. For centuries the concept of "catastrophism" in earth's history was generally accepted, with the great biblical flood as the basis. French scientist Georges Cuvier enlarged this to many catastrophes in an attempt to justify his belief in creationism rather than any kind of evolution or progressive change in living things. According to Cuvier, and others, catastrophes struck in various forms from time to time, so that life had to begin anew.

In 1785 Scottish naturalist James Hutton published his *Theory of the Earth,* in which he proposed that natural forces such as weathering, erosion, river channeling, and the like had been proceeding at a uniform rate for ages. This concept appealed to Britisher Charles Lyell, who wrote the important book, *Principles of Geology.* This coming of geological science, and the development of methods for estimating the age of the earth, made it evident that earth must be much older than 6,000 years.

The name "uniformitarian" was applied to the new geological theory by its opponents, and this theory is now generally accepted for the aging of the earth. The next logical step was to work backward through time to the beginning of the process. Knowing how much sediment was laid down each year by rivers, geologists measured the depth of accumulated sediment and began estimating earth's age at 100 million years or more. The effect on believers in creationism was explosive, and a battle began. It continues today, although with muted force and outside the concern of those who are more taken up with everyday matters than with when the world began.

Taking a page from the catastrophists' book, Georges Buffon suggested that the earth was formed when the sun collided with a comet. It was soon learned that such a collision would not be very violent, and this theory was dropped. However, Isaac Newton later suggested that the entire solar system began when a cloud of cosmic gas and dust collected, condensed, and collapsed to form the sun and its satellites.

This is still accepted as pretty much what happened, although Newton's picture was not complete or correct in every detail. Many brilliant men, including Immanuel Kant, Pierre Simon de Laplace, Hermann von Helmholtz, Lord Kelvin, James Clerk Maxwell, and Sir James Jeans added their suggestions. Not until 1944 did German astronomer C. F. von Weizsäcker propose his "eddy" theory, which refined Newton's cloud theory to the point that it seems a proper explanation.

Charles Darwin greeted the suspected antiquity of the earth with relief and appreciation, for it gave more time for natural selection to have worked. Since then, of course, the earth has been recalculated to be older and older until at present it is generally thought to be about 4.6 billion years old. Lunar materials seem to be about the same age and so strengthen the estimate for earth.

Dating a planet is a complicated and iffy process at best, not quite like reading the dates on old newspapers. Such exotics as iodine-xenon clocks are among the current techniques joining carbon-14 dating, uranium clocks, and other sophisticated methods. Carbon-14 dating is presently thought to be somewhat inaccurate, but geologists nevertheless feel they are correct about the age assigned to the earth.

Life is probably not as old as the formation of the hot cloud of gas that would become our planet, unless it was something quite different from what we are familiar with. It is doubtful that even the weirdest kind of evolution could account for creatures existing at a temperature of thousands of degrees. The coming of life must have had to wait for at least some minimum period and cannot span the full 4.6 billion years. Recent theories from Caltech suggest that the earth needed only about 10,000 years to solidify from hot gases and condense the elements that made up its core and crust. However, life still could not have come along immediately, for it is a long way, evolutionwise, from inorganic matter.

When Pasteur and others torpedoed the theory of spontaneous generation they also did away with the idea of the origin of new life where none had been before. Pasteur publicly stated that this was exactly what he intended to do—life was *created* once and for all by God in the beginning, and that was that. By Pasteur's definition, then, life had always existed. There were many others who believed this, including German chemist Justus von Liebig, who asked why organic life could not be just as much without a beginning as carbon apparently is. Another German, Wilhelm Preyer, agreed, and suggested that all we had to do was forget the notion that life must always have been in the forms we now find. He saw the entire molten earth as a living thing. That life survived the fire, shed the lifeless slag, and became the familiar protoplasm that exists today. Later the German biologist Eduard Pflüger suggested somewhat the same idea, basing primordial life on cyanogen, a gas made up of carbon and nitrogen.

Most scientists couldn't buy these ideas of red-hot life, of course, and theorized ways in which life could have begun after things cooled down to a reasonable temperature. Among these originists was Frenchman Jules Michelet, who more than a century ago suggested that life began in the ocean as drops of protoplasmic jelly swimming in water rich in nitrogen. From these living blobs, insects developed within about 10,000 years. After another 100,000 years man came along. Pasteur considered this heresy, and said so publicly. But in England Charles Darwin was considering evolution. There were two strikes against the idea of life originating suddenly from nonlife, of course. First, Pasteur had killed the notion of spontaneous generation. Moreover, if a simple life form *did* appear it would most likely be finished off—before it could take a deep breath—by all the living things that now pervade land, sea, and air. But, suggested Darwin cautiously, suppose we had a warm little pond filled with just the right kind of ammonia and phosphoric salts, with light, heat, and electricity in the environment? This concept remains today as the nearest thing to an acceptable evolutionary theory of the origin of life. It was worked out in some detail by the Russian A. I. Oparin, and published in 1938 as *The Origin of Life.*

According to Oparin, amino acids and proteins formed spontaneously. Then these molecules slowly aggregated into "colloids," or gelatinous groupings. After hundreds of millions of years, such aggregates with the property of cells developed. This happened when droplets were surrounded by a filmlike skin something like the membrane on cells as we know them. Nutrients permeated the film and caused growth that would eventually break the skin. Here, according to Oparin, was crude "reproduction," since the droplets contained thoroughly mixed components and the new cell was capable of sustaining "life" by itself.

Very cautiously Oparin suggested that the process of creation was so long and involved that it would be impossible to duplicate these forerunners of life in the laboratory. However, as time went on, more optimistic views prevailed.

Nobel Prize winner Harold Urey agreed with Oparin about the probable environment at the time life originated, and in 1952 a student of his, Stanley Miller, did an interesting experiment. For a week Miller bombarded a tiny environment of methane, ammonia, and hydrogen with electrical discharges. He reported the resultant formation of the amino acids glycine and alanine, plus the suggestion of others. The amino acids are the building blocks of protein. Also formed were formaldehyde, acetic acid, and succinic acid.

Following Miller's experiments, Melvin Calvin did similar "life-creating" experiments using a carbon dioxide environment containing water and hydrogen. High-energy radiation reportedly produced formaldehyde, acetic acid, and finally, succinic acid. This was a less successful result than Miller's and suggested that the methane environment was more likely for early life on earth.

As early as the 1940s English scientist J. D. Bernal theorized that life could not have originated in the ocean "soup" as suggested by Oparin. Instead, Bernal said, a denser material than water was indicated. His choice was clay. Interestingly, in 1969 Israeli scientists reported at the Third International Biophysics Congress that experiments at the Weitzmann Institute of Science had used montmorillonite, a common clay, in successfully linking amino acids into protein chains. Canada's A. G. Cairns-Smith is also interested in the clay-genesis theory, and suggests crystal formation as a starting point.

The Origin Gap

To date no one has taken the simple compounds created in these experiments and further developed them into proteins. For many scientists the concept of lucky accidents slowly converting organic building blocks into amino acids, proteins, cells, and finally more complex life forms seems fantastic. Indeed, some have "proved" that chance could never lead to life as we know it. Professor C. H. Waddington has likened chance development to throwing bricks into a heap in the hope that they will arrange themselves into a habitable house. The following admission was made by Oparin in his book:

. . . If the reader were asked to consider the probability that in the midst of inorganic matter a large factory with smoke stacks, pipes, boilers, machines, ventilators, etc. suddenly sprang into existence by some natural process, let us say a volcanic eruption, this would be taken at best for a silly joke. Yet, even the simplest microorganism has a more complex structure than any factory, and therefore its fortuitous creation is very much less probable.

The creation of a factory as described above, *even in a billion years,* would still be incredible. How, then, did even "more complex structures" arise fortuitously, with mere chance as the driving force? The oldest fossils found apparently date back about half a billion years. Fossils of course are far removed in the chronological chain of development from amino acids, and it is thought that the first primitive forms suggested by Oparin might have "assembled themselves" about 2.5 billion years ago. This leaves 2 billion years for the evolution from quasi-life to present-day human beings, a jump difficult to comprehend, since Oparin assigned hundreds of millions of years to the development of even a simple pseudocell. This implies something like half a billion years at least, leaving only 1.5 billion for evolutionary development from a subamoeba to an Einstein. Once nature got going she must have improved by leaps and bounds.

Creatures from Outer Space?

There are a variety of theories, old as the hills but ever new, on life from outer space. There was the "cosmozoic" theory, advanced by the nineteenth-century German chemist Hieronymus Theodor Richter, who included the discovery of the element indium among his more tangible accomplishments. This cosmic argument held that life came to earth from the far reaches of the universe, with spores wafted across the vast vacuum of space by stellar radiation. Hermann von Helmholtz, a brilliant scientist, subscribed to the similar "panspermia" idea, and suggested that meteorites could have brought life to earth. In 1908 Svante Arrhenius, one of the few Swedish Nobel Prize winners, proposed that life reached earth as spores driven by sunlight at 186,000 miles a second, somehow surviving the cold and the energetic radiation of outer space. In 1932, and again in the 1960s, scientists claimed to have found bacterial life in meteorites, although it was decided that their samples may have become contaminated with earthly bacteria.

Accepting the cosmozoic theory does not solve the creation problem, of course, but merely shifts the origin of life to a more distant setting. There remain two choices: either life has always existed (whatever "always" means) or it must have originated or been created at some point in past time. We may take our choice. It is easy to find fault with evolution theory, much easier than it is to produce scientifically satisfactory alternative explanations. Thus it is refreshing to hear the words of another Russian evolutionist of practical bent. V. Omeljanski dismissed the problem in this breezy manner: "When the Earth cooled off and the existence of organic life became possible, it appeared as the result of an unknown combination of matter and energy."

Natural Selection

Two generations before Charles Darwin, an earlier scientist of that same surname was making a beginning in evolution theory. Erasmus Darwin, who was a versatile inventor, expressed his theory in a poem called "The Botanic Garden." The crucial couplet was this:

> Hence without parent by spontaneous birth
> Rose the first specks of animated earth. . . .

Here was spontaneous generation of another sort, the first blending of inanimate elements into the first living assemblage of matter. This was an early, land-based version of the idea later expanded by Oparin.

Erasmus Darwin also believed that the use or disuse of an organ or other body part could result in changes which might be passed on to offspring. Buffon had earlier suggested the same method of change in creatures. The earlier Darwin thus did not move as far as a theory of natural selection to explain evolution. This step would wait for his grandson, Charles. Instead, Erasmus saw natural forces limiting population only by starvation and other catastrophes. Thomas Malthus seized on this idea and produced his theory of a population explosion that would doom humankind—a problem with us in greater force today, since the earth harbors three and a half times as many people as it did in Malthus's time.

Chevalier de Lamarck, born Jean Baptiste de Monet, was a military man turned naturalist and science writer for the masses. In 1799 he became professor of botany in Paris, although he was without much formal training. Lamarck was convinced, as was Erasmus Darwin, not only that characteristics could be acquired in animals and men, but that they could be transmitted to offspring. When a giraffe was donated to the Museum of Natural History in Paris, Lamarck pointed to it as the perfect example of what he was driving at: the animals had had to stretch to reach the leaves of tall trees, and thus each generation grew a longer neck.

Attractive as Lamarck's theory was, it elicited scorn from most scientists, particularly from biologist Georges Cuvier, who held the chair of anatomy in Paris. It was said of Cuvier that he could reconstruct any animal, given only a single bone. When a student prankster in devil's costume invaded Cuvier's bedroom and told him he was going to eat him alive, Cuvier reportedly yawned, replied that creatures with horns and cloven hooves were all vegetarians, and went back to sleep. Violently opposed to evolution of any kind, Cuvier savagely attacked Lamarck even after the man had died, and when asked to prepare a eulogy, wrote such a blistering denunciation that the Academy of Science refused to publish it.

At about this time Goethe proposed a romantic theory of evolution to the effect that all species were modifications of original "ideal" types. Goethe's ideal human, for example, was the ancient Greek. In a way he did not expect, Goethe's theory led to investigations showing that indeed there were gradations between species rather than the distinct different forms that had long been thought to exist. He also studied embryos and created the science of morphology (he coined the word), or how forms change. Cuvier also blasted Goethe and his followers, arguing from his lofty perch for a once-and-forever separation of living types with no crossing over of species boundaries possible.

As already noted, Charles Lyell's *Principles of Geology* completely rejected the catastrophism of Cuvier, and suggested a very great age for the earth. A pleased reader of Lyell's work was Charles Darwin. He had been grappling with an evolution theory, and Lyell's thoughts on the probable lifetime of earth and living things provided a much-needed assist. Darwin gave Lyell credit for "half my own writings," although Lyell himself, ironically, was firmly against the idea of evolution in living things.

As most of us have been taught, it was Darwin who proposed the idea of "natural selection" for the slow change of living species, and the creation of new species from old. *The Origin of Species* was published in 1859. But in 1831 a book called *On Naval Timber,* written by an obscure botanist named Patrick Matthew, clearly set out the basis of Darwin's theory. Of so little importance to the author that he relegated it to an appendix, here is his crucial statement:

This principle is in constant action, it regulates the colour, the figure, the capacities, and instincts; those of individuals of each species, whose colour and covering are best suited to concealment or protection from enemies, or defence from vicissitude and inclemencies of climate, whose fiture is best accommodated to health, strength, defence, and support. . . .

So little response did this "principle" evoke that Darwin learned of it only years after his own book was published, even though Matthew had written more than a quarter of a century earlier.

It is better known that just as Darwin was about to publish his great "discovery" he learned that another naturalist, Alfred Russel Wallace, had independently developed the same theory. Wallace was a self-taught naturalist who did his work in the jungles of Brazil and in the Malay Archipelago. He grasped the natural selection idea in 1857, and wrote a letter to Darwin in 1858, just before *The Origin of Species* went to press. As Darwin himself wrote, ". . . if Wallace had my MS sketch written in 1842, he could not have made a better short abstract!"

Darwin began to formulate his theory during a voyage on the British research vessel H.M.S. *Beagle* which sailed in waters including the Galápagos Islands. Here the young naturalist found species differing from island to island, growing in isolation and under different environments. He hypothesized that those individuals who chanced to differ in a way that better fitted them to the environment would reproduce more of their kind and in time predominate. "Darwin's finches" are given as classic examples of selection, fourteen variations which apparently evolved from one original species, each specialized to better fit the environment of its island.

"Darwinism" exploded like a bomb on scientists of that time. In weeks the battle lines had been drawn between creationists and evolutionists; between Darwinists and Lamarckians. One critic pointed out that since it was then believed that each parent contributed about equally in a sort of blending of characteristics, a new mutation would quickly be diluted to half, one-fourth, one-eighth, and rapidly vanish! Darwin, in all the decades of evolving his theory, had apparently not thought of this flaw, and began to suggest the possible operation of a kind of Lamarckism in later editions.

Great has been the ridicule heaped on Lamarck, even by current writers. Isaac Asimov, for example, dismisses the theory of acquired characteristics by pointing out that generations of circumcision have not altered the physiology of Jewish males. Darwin himself was not so sure, as the following account by him indicates:

Circumcision is practiced by Mohammedans, but at a much later age than by Jews; and Riedel, assistant resident in North Celebes, writes to me that the boys there go naked until from six to ten years old; and he has observed that many of them, though not all, have their prepuces much reduced in length, and this he attributes to the inherited effects of the operation.

Darwin also mentioned that a cat of his had its tail cut off and that the next batch of kittens was born with the same shortcoming! Most unscientifically he guessed at some sort of bodies in the bloodstream as being able to acquire changes during lifetime, and transmit these to the next generation. This subtheory he labeled "pangenesis." According to this hypothesis every part of the body prepares a kind of diminutive model of itself which Darwin called a "pangene." These pangenes are transported by the blood to the sex glands, where they unite to form the sex cells. The body part which has acquired a new character—say, an arm the muscles of which are strengthened by exercise—will bud off an altered pangene, and this pangene will produce a modified organ in the progeny. In a letter to his botanist friend J. D. Hooker in 1871, Darwin wrote, "I am always delighted to see a word in favour of Pangenesis, which some day, I believe, will have a resurrection."

Darwin also once wrote to Alfred Russel Wallace that critics had convinced him that his theory of natural selection was wrong, and some writers suggest that Darwin may have died still thinking that. Not so the German naturalist Ernst Haeckel, who read and completely accepted *The Origin of Species*. In fact, he proceeded to extend it dramatically with his own theory that the embryo of a living thing entirely "recapitulates" evolution as it grows. This belief that "ontogeny recapitulates phylogeny" made great headway for a while, is still accepted by some, and is taken as partly true by many more.

Haeckel also plunged in where Darwin had the tact to remain silent: he drew up the "family tree" of man, graphically showing the descent of man that Darwin had only hinted. Starting with the protozoa, Haeckel traced life through all creatures and ultimately to apes and men. To complete the roots of the tree, he suggested that the most primitive living thing had evolved from inorganic material.

Since Darwin's original intuitive guesses into evolution, Mendelian genetics has provided a mechanism to bolster it. However, botanist Hugo De Vries and others suggested gross mutations as the sole source of genetic change, and for a time this seemed the answer. Darwinism was exploited in one direction as "Social Darwinism," which tended toward racism. More recently it has become "neo-Darwinism," natural selection theory updated by later findings, and made more flexible to take in observed discrepancies.

Chance and Change

Theories of evolution depend on the idea that there have been changes in life forms. If this is not true, it would seem that the creationist is correct, that the Creator at once put everything into motion in a week or whatever time period was required and that everything then stayed pretty much the same. This may actually be the easier comprehended choice. Jacques Monod, French molecular biologist who shared a Nobel Prize in 1965 for his work on the genetic code, estimates that life has existed for about 3 billion years of the earth's age. Thus, it took more than 1.5 billion years for it to happen in the first place. (Monod, interestingly, believes there may have been just one chance for life, bringing to mind the remark about life being one of the more amusing properties of carbon.)

Bacteria, Monod estimates, appeared about a billion years ago. Thus 2 billion years elapsed between whatever was first and what today is the simplest form of life on earth. From that time to the differentiation of at least the main phyla of animals required only another half a billion years. Since that time, according to Monod, some living things have not changed appreciably. He mentions *Lingula* (marine invertebrates) as examples which have remained static for some 450 million years. Oysters 150 million years ago probably had the same appearance and flavor as those today.

According to Monod, hundreds of billions of mutations, or genetic changes, actually do occur to humans in each generation simply through accidents of one sort or another. This is so because of the number of people and the complexity of human makeup, consisting of tens of trillions of cells with perhaps 150,000 genes per cell. However, even though a human being is subject to genetic change, the chance of this being a transmissible change is very small. Most of the genes in our bodies are not in the sex cells we will contribute to our offspring, of course. Those that are number in the thousands or millions. Suppose that ionizing radiation does reach the sex cells and alters the actual gene structure. In most cases this will require such a massive dose as to make the mutation genetically lethal, or perhaps even destroy the entire organism so that it cannot reproduce.

In the species called *Homo sapiens* there exists a tremendous genetic variation potential. While we think of DNA as invariant, and reproducing accurately for hundreds of millions of years, there is nevertheless a broad range over which we still call the result human. There are white people and black people, and numerous shades between. There are many body types, blood types, hair colors, bone structures, and even organ arrangements (such as a single kidney, or several kidneys). Six fingers do not make one nonhuman (two heads would make a freak, but likely result in death before it could produce more two-headed humans).

Within the genetic program there is thus a rich potential of billions of different human results; variety is there in plenty without any genetic accidents. Suggested therein is the technique of eugenics, pioneered as a concept by Francis Galton but yet to make any appreciable impact on the human species. This is the theory of breeding selectively to produce humans who are "better" by some sort of yardstick. Plant and animal breeders have of course been highly successful in creating more productive crops and domestic animals of many types. This is the eugenics principle applied to lower types of life, and demonstrating clearly that it can work. These breeding operations are not the same thing as genetic mutation, however. Only in a limited segment of plant experiments has actual genetic change been accomplished. Using controlled nuclear radiation, scientists have succeeded in creating better peanuts, and one or two other kinds of crop plants. In more drastic experiments with animals, the screwworm fly has been all but eliminated as a species in some parts of the world by producing sterile males and introducing them into the insect population by the millions, where they mated with normal females and thus brought the line to an end. (However, as ranchers in the U.S. and Mexico learned in 1973, even sterilization hasn't completely rid them of the screwworm menace.)

Darwin said that natural selection accounted for the coming of new species, that it caused all the hundreds of thousands of different forms of life. Thus it must also have accounted for the jump from plant to animal life. Yet the creation of species seems to have come to an end long ago, for evolutionists can bring forth only questionable examples of new species. The only evolutionary change at the species level seems to be the dying out of some species. It is commonly stated, for example, that most of the species that ever lived have fallen by the natural selection wayside and vanished from the scene.

If it was once easy for life to cross the species line, the same conditions do not seem to prevail any longer. An example is the impossibility of the interbreeding of different species, except in very rare cases, such as the horse and the donkey, which are of course quite similar to begin with. Even here, there is enough of a barrier that the resulting hybrid offspring, the mule, is sterile. Buffalo and cattle also can interbreed, and lions with tigers. From time to time pranksters claim to have interbred other species. For example, an Englishman recently claimed to have mated a dog and a cat and produced a litter of "dogats." Such hybrids, if authentic, would have brought a pretty price, not only from science but from pet lovers as well.

Biologists have announced that the DNA of mice is much the same stuff as the DNA of human beings, a seemingly humbling admission. However, there must be more to the genetic mechanism than that, for men still produce men and mice mice, and the question of whether one is man or mouse remains rhetorical. Human beings have been known to copulate with animals, including those as ordinary as the barnyard variety and as exotic as the manatee (which probably led to the myth of the mermaid). However, despite cautionary tales of animals giving birth to young with human faces or other attributes cited by Puritans in punishing offenders, man probably has not crossed the species barrier either.

Still Some Missing Links

It seems probable that, as Russian biologist V. Omeljanski breezily put it, when things were right on earth, life originated. Likewise, despite the difficulties with theories, life probably does change in some form of evolution. Darwin's theory of natural selection is as close as any scientist has come to explaining what has happened since then, and supposedly continues to happen. However, it should not be assumed that all the problems are solved and that there is general agreement by everyone but the creationists that Darwin said it all in 1859. The truth is far from that. A case in point is suggested in the recent controversial book by Arthur Koestler, a writer with more than a nodding acquaintance with science. *The Case of the Midwife Toad* describes the neo-Lamarckian theory of Paul Kammerer, an Austrian naturalist who bred amphibians, including the midwife toad, and was convinced that environment *did* cause changes in them, and very quickly at that. In a somewhat biased account, Koestler documents much unscientific action on the part of neo-Darwinist biologists, and some interesting findings by Kammerer, who committed suicide shortly after serious doubt was cast on the authenticity of his experimental work.

Harvard zoologist Stephen Gould argues that Koestler has completely missed the point of neo-Darwinism but admits that Darwin-

ists have been "singularly unsuccessful in conveying our understanding of natural selection to interested nonscientists." Gould
points out that Kammerer succeeded only in exploiting genetic
potential latent in his experimental subjects—they did not acquire
the controversial "nuptial pads" from scratch but simply called on
them from gene potential.

Three years earlier in *Science,* the journal in which Gould
corrected Koestler, geneticists Jack Lester King and Thomas H.
Jukes argued for "non-Darwinian evolution" through neutral mutations, rather than those selectable for desirable characteristics.
Included in their paper was the suggestion: "The idea of selectively
neutral change at the molecular level has not been readily accepted
by many classical evolutionists, perhaps because of the pervasive-
ness of Darwinian thought." According to these non-Darwinists,
"Evolutionary change is not imposed on DNA from without; it
arises from within. Natural selection is the editor, rather than the
composer, of the genetic message. One thing the editor does *not*
do is to remove changes which it is unable to perceive."

Although neo-Darwinist natural selection is accepted by most
scientists as a fact of evolutionary life, there is an articulate minor-
ity who feel that a question put forth by Alfred Russel Wallace to
Darwin has never been satisfactorily answered. Wallace's question
was simply this: If evolution comes through natural selection, a bit
at a time, how is it that man seemingly sprang into the world
mentally and physically endowed to do far more than he was called
upon to do in his dawn years? How, in only one *million* years, did
the human nervous system (including a brain that can speculate on
itself to the extent evident in the diligent quest for an evolutionary
theory!) evolve by blind chance? Wallace put it this way:

How then was an organ developed so far beyond the needs of its
possessor? Natural selection could only have endowed the savage with
a brain a little superior to that of an ape. Whereas he actually possesses
one but little inferior to the average member of one of our learned societies. . . .

An instrument has been developed in advance of the needs of its possessor.

Darwin couldn't go along with this concept at all and told Wallace that he differed grievously from him and was sorry for it. However, he was not able to answer Wallace's seemingly valid question. As Loren Eiseley says in *The Immense Journey:*

Slowly Wallace's challenge was forgotten and a great complacency settled upon the scientific world. . . .

Ironically enough, science, which can show us the flints and the broken skulls of our dead fathers, has yet to explain how we have come so far so fast, nor has it any completely satisfactory answer to the question asked by Wallace so long ago.

1. If differences in mental abilities are inherited, and
2. if success requires those abilities, and
3. if earnings and prestige depend on success,
4. then social standing will be based to some extent on inherited differences among people.

Richard Herrnstein, *Atlantic Monthly,*
September, 1971

4   Nature and Nurture

Richard Herrnstein in 1971 pointed to well-established evidence
that intelligence is heritable, and both shocked and angered many
with the "syllogism" above. His conclusion came as no surprise to
"hereditarians"—and perhaps to many laymen as well—but was a
bitter pill for "environmentalists" to swallow, and many refuse to
do so. Thus an ancient controversy erupted like a newly awakened
volcano to liven up the current genetic revolution.

For centuries the lowliest peasant knew that you "can't make a
silk purse out of a sow's ear." If this is true, it seemed to follow
logically that neither could a person born without the necessary
mental equipment be a genius. This thesis was formally pro-
pounded a century ago by Francis Galton, an Englishman who

amply demonstrated that he was one of the hereditary "haves" in the mental department. When his book *Hereditary Genius* was published in 1869, another brainy Englishman named Charles Darwin wrote to the author: "I do not think I ever in all my life read anything more interesting and original." The two men were cousins, a biological fact that may or may not lend credence to the theory of inherited intelligence.

For Galton, heredity was quite simple. He had inherited longevity from his family; he also noted that physical attributes were largely inherited. Not surprisingly, his research indicated to him that mental capacity as well was to a large extent hereditary. Just as there was a "normal distribution curve" for height, weight, and other measurements, there was one for intelligence. A few geniuses are born in each generation, and a few idiots, with the bulk of the population clustered around an "average intelligence" later to be pegged at 100 in IQ tests.

Even a century ago the "nature versus nurture" battle had raged for a long time. Galton's work was largely a reaction to the idealistic belief of many environmentalists including Rousseau and Locke that proper surroundings could make superlative beings of us all. Early in his book Galton stated flatly that he had no patience with those who, in "tales written to teach children to be good," held that all babies were born pretty much alike. Instead, he claimed there was more mental difference between the brightest and dullest humans than between a normal human and a normal dog! Galton set down a simple definition of what he meant by hereditary endowment:

By natural ability I mean a nature, which when left to itself will, urged by an inherent stimulus, climb the path that leads to eminence; that has strength to reach the summit—one which, if hindered or thwarted will fret and strive until the hindrance is overcome and it is again free to follow its labor-loving instinct.

In staunchly supporting his theory of heredity, Galton nevertheless maintained the scientific balance to admit: "Man is so educable an animal that it is difficult to distinguish between that part of his character which has been acquired and that which was in the original grain of his constitution." His use of the word "grain" is interesting. When *Hereditary Genius* was published there was not yet available the theory of genetics that Mendel's work would lead to by the end of the century. With his cousin Charles Darwin, Galton was groping with externals, treating the hereditary mechanism as a biological "black box" and inferring what must go on inside it.

In *Science for the Citizen,* Lancelot Hogben makes the point that Jean Jacques Rousseau also used commendable restraint in not going beyond the biology of his time. Indeed, in his essay "The Origin of Inequality" Rousseau wrote: "I conceive that there are two kinds of inequality among the human species: one, which I call natural or physical, because it is *established by nature* and consists in a difference of age, health, bodily strength, and the qualities of the mind or of the soul; and another, which may be called moral or political inequality."

Despite this fair basic premise, however, some of the French philosopher's disciples espoused views which gave too much emphasis to environment and too little to heredity. Rousseau himself seems to have done this in later writings which clearly suggested that good environment would solve all problems. The resulting humanitarian approach extended to attempts at rehabilitation of the mentally deranged in England and America, "generally inspired by a philanthropic zeal which drew little inspiration from scientific knowledge," according to Hogben. When such good works failed to help the mentally ill and socially deprived there was a reactionary swing in the other direction, and Social Darwinism had its innings.

Traditional belief in the "great chain of being," a pecking order rigorously fixed on us by nature, made comfortable those who believed that kings were born kings and peasants peasants, and never the twain could meet. Eugenics, a term coined by Galton and taken up by the Social Darwinists, held that two things were necessary to produce the great society of man. First, the bright should marry the bright, and government should take steps to expedite this breeding. Second, the dregs of humanity should not be allowed to propagate more unfortunates who obviously could not be helped even by the best-intentioned social workers.

Hogben described the resulting nature-nurture squabble with clarity and humor when he wrote: "Educational reformers with radical views often justify them by arguments which suggest that caterpillars of the cabbage butterfly will take to a mixture of pollen and honey. Their opponents appear to hold that Newton would have written his *Principia* if he had been born in Tasmania."

American behaviorist John Broadus Watson in the 1920s added fuel to the environmentalist fire by stating in his book *Behaviorism:*

Give me a dozen healthy infants, well-formed, and my own specified world to bring them up in and I'll guarantee to take any one at random and train him to become any type of specialist I might select—doctor, lawyer, artist, merchant-chief and yes, even beggarman and thief, regardless of his talents, penchants, tendencies, abilities, vocations, and race of his ancestors.

It is not recorded that Watson ever made good his guarantee; there is serious question that he meant it literally or expected to be taken at his word. However, the lines were still drawn as sharply as ever between opposing camps. Time has done little to patch up the basic differences between biologist and sociologist. In 1949, in *The Science of Culture,* sociologist Leslie White wrote: "From the standpoint of human behavior . . . all evidence points to an utter insignificance of biological factors as compared with culture in any consideration of behavior variations."

Three years later, geneticist C. D. Darlington, in *The Facts of Life,* disagreed completely: "Owing to inborn characters we live in different worlds even though we live side by side. We see the world through different eyes, even the part of it that we see in common.

. . . The materials of heredity contained in the chromosomes are the solid stuff which ultimately determines the course of history."

To environmentalists, this seemed a facetious argument that it is not life itself that is of greatest importance, but nucleic acid, the genetic material perpetuated through the ages from one individual to another and one species to another.

Countering hereditarian claims, environmentalists point to indications that proper training can change mental retardees into normal and even above average individuals. An example is work done at the University of Wisconsin by psychologists Rick Heber and Howard Garber with children of retarded mothers. Selecting forty mothers and newborn offspring, Heber and Garber assigned half at random to a control group which was left to function normally, with retarded mothers caring for their babies at home and by themselves. The other half were cared for by "infant stimulation teachers" at the university during the day. The results claimed are remarkable.

Years ago psychologist Lewis Terman in his study of California geniuses reported that his subjects, with average IQ's of about 150, produced children averaging about 130. This was accomplished in a setting of successful marriages, a very low divorce rate, success in business or profession, and a generally very good environment. But the Wisconsin researchers claimed in just five years to have raised the IQ of the babies of retarded mothers (who averaged 75 IQ or less) to 125 on the average. Since control babies averaged only 95 IQ, environment was claimed to add 30 points to the intelligence of the children. An indicated experiment would seem to be to give the retarded mothers themselves similar treatment to bring them at least to average intelligence.

B. F. Skinner is perhaps the most vocal of behaviorists, still promising the kind of conditioning that Watson talked about half a century ago. In *Beyond Freedom and Dignity* Skinner sees no need for genetics, relying on environmental conditioning that will mold human beings to fit the world of the behaviorists—although he frankly admits that most will not like that kind of world at the outset.

Recently Dr. Saleem A. Shah, chief of the Center for Studies of Crime and Delinquency of the National Institute for Mental Health, voiced the opinion that even though the nature-nurture debate is seldom voiced in the "polarized and acrimonious tones of a few decades ago," recent pitched battles between geneticists and behaviorists indicate that if anything the debate is fiercer and hotter. Specific examples are the violent attacks on Arthur Jensen, William Shockley, and others who suggest that tests indicate a difference in intelligence among races and that therefore different kinds of education are needed. It is obvious that the debate has also become politicized, or ideologized to a great extent. In 1949 a survey of twenty-four psychologists, biologists, and sociologists showed them already split almost totally along party lines: of twelve political liberals or radicals, eleven were environmentalists; of twelve political conservatives, eleven believed hereditary factors were crucial.

The IQ Argument

In the book *Heredity, Race and Society,* L. C. Dunn and Theodosius Dobzhansky noted: "Seventy percent of Americans can taste a weak solution of PTC, which is phenothyocarbomide. They taste this as intensely bitter. To the other 30 percent it is almost tasteless." Here is a genetic ability—or disability. No amount of will power can make a PTC taster of a nontaster, or vice versa. No one with red hair, or no hair at all, can will himself a mop of blond hair. There is little fighting going on over these hereditary facts of life. The stickiness generally comes with the subject of intelligence.

Human beings are made up of cells, each of which incorporates a genetic mechanism regulating its growth and development. The body, *in toto,* seems hereditarily determined as to size, weight, shape, color, and so on. We are also physically subject to environmental forces, of course. If a safe falls on a man, his chances of prospering, or even of continuing to breathe, are materially reduced if not eliminated entirely. Should a person eat more he will probably weigh more; subject to an environment of coal dust, bright sunshine, or corrosive gases, skin color and texture may change accordingly. But this does not change the fact that there are genetic limits too.

The central nervous system is likewise cellular in nature. The brain is made up of several billion neurons, also endowed with built-in genetic controls. There is a critical difference between nervous system cells (and the cells of some organs) and other cells: nervous system cells cannot duplicate themselves, and are constantly being reduced in number. No matter how much fish or other "brain food" we eat, we do not increase the number of brain cells. In the mental department, then, we are even more limited by our hereditary endowment.

Original Lamarckism was based on the commendable notion that "will power" could make us better, and this thesis was taken to heart by sincere humanitarians. Dr. Émile Coué put it into a little jingle, religiously repeated by many of an earlier generation: "Every day, in every way, I am growing better and better!" Here was the "power of positive thinking," surely better than skulking around muttering that we are getting worse and worse. But the question the hereditarians ask is if *any* amount of practice can convert 97-pound weaklings into Olympic stars, or if any amount of positive thinking can upgrade an average mentality into an Einstein or Shakespeare.

The crux of the nature-nurture argument seems to be whether or not the human mind is truly a blank slate at birth, as Locke and others argued to set the stage for the environmentalists and egalitarians, or whether Kant and those of his persuasion were correct in their belief that such a concept is like expecting a random heap of cards in a library to somehow neatly arrange themselves into the Dewey decimal system.

The hereditarian argues that there is a certain amount of pre-wiring in the human brain—in any brain for that matter—and that this genetic head start or handicap determines the geniuses and the dullards of any population. Thus heredity determines who can be a mathematician, who can best play chess, and who can be artistic in temperament. While certainly environment has an effect on our development, and not many Miltons flourish in isolation or depriva-tion, some are destined to do better and some to do worse, under given conditions.

Galton suggested that an excellent way to test the influence of heredity on intelligence was to measure the intelligence of identical twins. This has been done to some extent, and the results are in-teresting. In his book *The Nature of Human Intelligence,* J. P. Guilford cites 56 studies, involving 113 groups of variously paired human subjects, ranging from one-egg twins reared together to unrelated pairs reared apart. Fifteen of these pairs were identical twins raised together, and thus exposed to a similar environment. Their intelligence correlated by a factor of .88 (1.00 would be identical). Four pairs of one-egg twins reared apart gave a correla-tion of .75. According to Guilford, the difference between .88 and .75 represents the effect of environment on the subjects.

Nonidentical twins, both in like-sex and unlike-sex pairs, showed a correlation of only .53, again a strong indicator of the power of heredity in determining intelligence. Unrelated children, reared together, showed a correlation of only .16; and unrelated, reared apart, only .09.

These studies seem to bear out the contention of hereditarians that genetics is involved in intelligence. Other studies, such as those by Dr. Lewis Terman of genius children in California, indicate that brilliant people produce children brighter on the average than those of less intelligent parents. Nevertheless, environmentalists generally tend to remain unconvinced, and even profess to see in such results proof of their own theses.

There is a mathematical formula for predicting the intelligence of children from that of their parents. First, the IQ's of their parents are averaged. Next, 100 is subtracted from that figure, and the remainder multiplied by 80 percent. The result, added to 100, gives the child's probable IQ. For example, if a father has an IQ of 145 and the mother 155, their average is 150. Subtract 100, leaving 50. Multiply this by 80 percent to get 40, which added to 100 equals 140, the predicted IQ.

This formula indicates some very interesting things about inherited intelligence. There is a tendency for the IQ of children to move toward the norm of 100, whether their parents are above average or below. Parents averaging 80 will produce children with IQ's of 84, on the average. Such improvement statistically would lead to increasingly bright children until they reach the norm of 100.

The coming of the Darwinian revolution produced great resistance in creationists. Today the nature-nurture controversy produces similar opposition in orthodox environmentalists—who form the majority of the social scientists.

Even those environmentalists who grudgingly admit that heredity has some slight effect on performance argue that IQ tests are invalid: whatever the tests measure, it is not total intelligence. This charge may well be true, for "natural ability" is not the simply identifiable thing that Galton hoped to find in his measurements of hearing, vision, and simple motor accomplishments. J. P. Guilford is among those suggesting that there are dozens of different kinds of intellect; he lists 81 in his book. Nevertheless, IQ tests demonstrably are measuring *something,* perhaps even a portion of Galton's "natural ability."

Sociologist Lionel Tiger, a Canadian now teaching at Rutgers, has suggested charging anyone administering an IQ test with statutory rape, a rather extreme suggestion. However, although IQ tests have a long history of quite good correlation between test scores and performance in the "real world," there have always been criticisms of their validity. Bias against minorities is often charged, for example. While testers claim to have eliminated such biases so that there are "culture-independent" tests, it seems obvious that there must always be some handicap for certain test takers, at certain times, and under certain conditions.

One who firmly believes this is Dr. John Ertl, now at the University of Ottawa. In 1959 Ertl took an IQ test during his graduate work in psychology and scored only 77, a level that indicated he was in the dull-normal category. Understandably perturbed, Ertl set out to find a more accurate method, and his results seem to indicate that in his case at least conventional IQ testing had serious shortcomings.

Brain-wave patterns have long been used to study mental disorders and Ertl decided that they might be adapted to measuring latent intelligence as well. Since 1965, he and other researchers have been reporting a fairly high correlation between brain-wave responses and intelligence. Commercial testing equipment is available, and the range of applications is surprising. Dr. Rudolph Engel of Oregon Medical School in Portland believes that the Ertl test is the most accurate measure of at-birth intelligence available. And in 1972 the U.S. Navy began a study on recruits to see how well brain-wave results correlated with demonstrated performance.

Far out as the brain-wave intelligence test seems, it is topped by another being investigated by a Scottsdale, Arizona, psychiatrist. In this test, the measurement of ears is crucial! Dating to the 1860s, when Italy's Cesare Lombroso tried to link various physical characteristics to congenital brain damage, the technique has recently been revived by Dr. Kent Durfee. By late 1972, Durfee had measured 274 children known to be abnormal and 133 who were normal. He found that 96 percent of mental retardees measured had asymmetrical ears. This same difference in ear shape prevailed for about 70 percent of emotionally disturbed, 78 percent of juvenile delinquents, and 79 percent of academic underachievers. However, only 29 percent of normal subjects showed the ear asymmetry. Durfee carefully notes that his results are tentative, and that much more research is needed.

It is questionable how well the medical world, educators, employers, and the public in general would take to such exotic "intelligence testing." Hinging one's career or that of one's children on a brain-wave pattern or a droopy ear sounds like an ultimate act of faith. Obviously much more field testing of such methods is in order. Meantime, conventional IQ tests, with whatever bias they contain, continue to be used in a variety of fields.

It has been demonstrated that a person with a low IQ *rarely* does well in school. One with a high IQ does not *always* do well, and in fact such people perform at all levels in school. The implication thus seems to be that intelligence (as measured by IQ tests) is *necessary* for success in school, but not a guarantee. Motivation, personality traits, and other factors must also be involved. Nevertheless, Lewis Terman in 1925 challenged those who thought IQ itself could be raised to prove their contention. It would cost only a few hundred thousand dollars, at most a few millions, he said, and the knowledge would be a thousand times as valuable to humanity as the cost. In 1969 Arthur Jensen stated that the expenditure of *billions* of dollars by the educational system to do just what Terman suggested nearly half a century earlier had completely failed to raise the intelligence of the subjects. According to Jensen, compensatory education was a failure because intelligence is largely a hereditary capacity and cannot be appreciably increased.

More recently, sociologist Christopher Jencks, who had already disquieted educators with his earlier book *The Academic Revolution* (with David Riesman), stated bluntly that education itself cannot be expected to improve the lot of a student who does not have genetic intelligence and a good home environment. His new book, based on studies funded by $750,000 worth of Carnegie Corporation of New York grants, is titled *Inequality: A Reassessment of the Effect of Family and Schooling in America.*

Intelligence and Sex

Of the 1,500 subjects in Professor Terman's California tests, 857 were boys and only 671 were girls. Proof of male mental superiority? Not necessarily. Women have traditionally been treated as second-class citizens in the brains department. Although Eve was first at the tree of knowledge, her daughters have received precious little credit, and not until recently were women accepted even as bright enough to vote. In spite of this sop, the stigma remains that the physically weaker sex is also weaker in mental ability: "Who ever heard of a really top-flight woman scientist, author, or whatever?" It is largely such unfair treatment that has goaded some of the more sensitive females into pushing for "women's liberation." Yet, with plenty of firm ground from which to mount the attack, many partisans have gone off half-cocked, in as unscientific an approach as could be imagined.

Just as the environmentalist claims there is no difference among human beings but that we are all created naturally equal, some women claim that only male chauvinism keeps them on the distaff side of life. Unisex, a synthesis any thinking human will reject if he (or she) is more than a vegetable, is pushed by extremists as an antidote to the perceived "androgenous society" of the male-dominated world. Actually, *la différence* the French have long celebrated goes far deeper than the outward physical attributes of sex, as a little thought indicates.

Women make better mothers than men because men cannot conceive and bear young. Women are undoubtedly handicapped during menstrual periods. Generally, they are not the equal of men in physical strength. Neither are they as subject to baldness, color blindness, or hemophilia. And, while nobody in his or her right mind claims that women are mentally inferior to males, it is a physiological fact that they are somewhat *different* in the brain department. All these are genetic differences.

The average female brain is somewhat less than 90 percent as large as a man's. But relatively this is about the same proportionate weight of brain to body. How important the absolute weight of the brain is remains unknown. Anatole France's brain weighed only about 1,000 grams; on that scale the average female brain of 1,585 grams should be ample for a genius and a half. In his book *The Differences Between a Man and a Woman,* however, Englishman Theo Lang continues to argue for the superiority of men over women in intelligence. There has been no female Beethoven, for example. Or Einstein, or Edison.

Actually, standard IQ tests show remarkably small overall differences between men and women. We are more alike in our minds than we are in other sexual characteristics. However, the IQ testers, beginning with Binet and continuing to the present, found that there were sex differences in the results of intelligence tests. This is not to say that men are smarter than women, or that the reverse is true either, but in tests each sex did show more ability in different areas. For example, female subjects demonstrate more ability (on the average) to manipulate symbols in coding operations and to recognize similarities between different objects. Males demonstrate better mathematical reasoning and spatial conceptualization, plus a wider store of information. Specific examples are the Witkin Embedded Figures Test, in which males score higher, and tests for naming in rapid sequence the colors of blocks in a pattern, in which women excel.

J. P. Guilford surveyed all available information on sex differences in intelligence tests and presented the following list in *The Nature of Human Intelligence:*

Males Higher	Females Higher
Street Gestalt Completion	PMA Reasoning
Spatial Orientation	Opposites, Verbal Analogies
Spatial Visualization	Wechsler Similarities
Porteus Maze	Memory for Figures
Arithmetic Reasoning	Digit Symbol
Match Problems	Memory for Words
Gottschaldt Figures	Word Fluency
	Ideational Fluency
	Expressional Fluency
	Symbol Identities

Environmentalists explain away male superiority in various categories on the ground that boys and men are exposed to more areas, they are involved with mechanical things, and so on. This may be true, but are girls exposed to more memory work, reasoning, words, and so on?

The miracle of sex determination in humans, resulting in just about one boy for every girl, and vice versa, is accomplished by an important characteristic of male and female sex cells. In conception, the accident of chance (a euphemism for a variety of complex factors) determines if the fertilized egg will contain one X and one Y chromosome, or two X's. XY results in a boy baby, XX a girl. Now, the basic human model is an all-purpose type, as a bit of speculation on male nipples will indicate. Useless to a man, these would have been needed had he been a woman.

Thus the basic human design also seems to be female. If, at a certain stage of fetal development, a spurt of the male hormone androgen is not triggered by the genes, the brain, along with other parts of the new organism, will remain female. If the hormone *is* forthcoming, the male sex apparatus appears, and more subtle changes occur in the tiny developing neural circuits of the brain. Dr. Leon Salzman of Georgetown University Medical School believes that while genetic sex is established at fertilization, the effect of the sex change is not felt until the fifth or sixth week of fetal life. Until that time, all fetuses are female.

Occasionally nature goes awry and produces a "genetic male" who is outwardly a female. This variation has internal testes in place of ovaries, and no uterus. Such mutations often grow up contentedly as women, although of course sterile. There is also the "genetic female" who receives *excessive* amounts of androgen or similar hormones. Cause for these anomalies may be congenital malfunction in the fetus, or perhaps hormone therapy given the mother during pregnancy.

The knowledge that hormone therapy during pregnancy can produce different sexual types leads to suggestions that this is the way toward the sexual equality some militants urge. What is done incidentally might be repeated and done better by intent. Why not inject women with androgen to make them more masculine—and inject men with female hormone to make them less dominant and aggressive? Why not perform chemical changes on all fetuses and produce a single-sexed society? There are those who applaud such an approach, whereas others would agree with Dr. Edward Zigler, director of the Federal Office of Child Development, who says, "Perhaps many of these genetically influenced behavioral differences could be overridden through training, but only at some considerable psychological cost. I think that such a thing as being true to one's self makes sense genetically as well as making for a much more interesting society."

Superbrains

Early in 1972, Dr. John Money of Johns Hopkins University School of Medicine and Hospital in Baltimore announced the finding that an excess of certain sex-related hormones in the fetus may significantly increase intelligence in the child. The hormones under investigation are androgen, the male sex hormone present in both males and females in varying amounts, and progesterone, the pregnancy hormone.

Earlier we discussed the genetic variants resulting from excess amounts of androgen in the fetus. Another congenital condition is known as the androgenital syndrome. Studies of seventeen persons (age two and a half to forty-eight) with this syndrome disclosed that a "remarkably high proportion" had high IQ's. Specifically, almost 13 percent had IQ's higher than 130 and more than 60 percent were higher than 110. Both these percentages are several times higher than normal for the entire population. Money pointed out that a direct link had *not* been proved between hormones and high intelligence, only the fact that both were present. However, he suggested that continuing research may shed light on the causes of both genius and the mentally deficient.

Other researchers feel that mental deficiency is caused in part by oxygen starvation in late pregnancy. Indeed, South African physicians have subjected volunteer pregnant women to a regimen of decompression during pregnancy which, in some opinions, has produced a higher percentage of above-normal-intelligence babies than would normally be expected.

Dr. Abraham Towbin of Harvard Medical School has written in the *Journal of the American Medical Association* that in effect all human beings are mentally retarded to a degree. This is caused by mechanical injury to the central nervous system, as well as oxygen starvation to the system. Particularly is this true, Towbin says, when birth is premature. Huxley had anticipated this prospect too, as the following dialogue from *Brave New World* demonstrates:

"Reducing the number of revolutions per minute," Mr. Foster explained. "The surrogate goes round slower; therefore passes through the lung at longer intervals; therefore gives the embryo less oxygen. Nothing like oxygen-shortage for keeping an embryo below par." Again he rubbed his hands.

"But why do you want to keep the embryo below par?" asked an ingenuous student.

"Ass!" said the Director, breaking a long silence. "Hasn't it occurred to you that an Epsilon embryo must have an Epsilon environment as well as an Epsilon heredity?"

It evidently hadn't occurred to him. He was covered with confusion. "The lower the caste," said Mr. Foster, "the shorter the oxygen." The first organ affected was the brain. After that the skeleton. At seventy per cent of normal oxygen you got dwarfs. At less than seventy eyeless monsters.

Many experiments have been done toward improving memory, and thus intelligence, since the two are thought to be correlated. A variety of drugs have been reported to improve memory and learning ability; glutamic acid was an early contender. Later a trade preparation called Cylert was credited with dramatic improvement in memory of older people. Other experiments seemed to indicate that doses of yeast RNA worked wonders for those with memory problems. However, positive results have yet to be documented for these techniques and the even more dramatic "learning transfer" experiments which seem to suggest that learning can be transferred as a chemical extract rather than a process.

Intelligence and Race

Tradition has conferred, and most probably mythically, superiority in athletics and sexual activities on blacks. Huxley's *Brave New World* engineers made a point of fecundity: ". . . Mombasa has actually touched the 17,000 [eggs from one ovary] mark. But then, they have unfair advantages. You should see the way a Negro ovary responds to pituitary. It's quite astonishing when you're used to working with European material."

Blacks are also, supposedly, the inheritors of inferior brains, a condition which would handily explain their low economic status. However, many say that the cart has been put before the horse in this explanation and it is the fact that they are disadvantaged that makes them score low in IQ tests. This argument certainly has merit.

Some researchers, notably including Audrey Shuey, Arthur Jensen, William Shockley, and H. J. Eysenck, believe there is documented proof that the Negro lags genetically in intelligence. However, those with this hypothesis have been all but ostracized from the social sciences. Shockley has been refused opportunities to speak at scientific conferences and on college and university campuses. Jensen has been savagely attacked by environmentalists, as has Eysenck, whose book on IQ is dismissed by some as ridiculous nonsense, despite the author's professional standing.

Basis for the race intelligence controversy seems to be the observation by several researchers that in North America Negroes score about 15 points lower in IQ tests than do whites. Jensen suggests that this cannot all be attributable to disadvantage, since other minorities—for example, American Indians, who may be more disadvantaged than blacks—score higher. Eysenck points out that Chinese minorities, probably still under some social handicap, score higher than white native Americans.

Reasons advanced for the supposed shortcomings of blacks range from general lower mental ability because of race to the supposition that lower mentalities would have been selected as choice slaves (a dubious hypothesis) and that the American Negro does not represent a cross section of African population.

Since Jensen is quick to add that the difference in IQ is only on an average basis, and that some blacks score in the genius range, it would seem that his suggestion of different educational techniques should apply to all those of low IQ, rather than single out a particular race.

"Natural Ability," and Where It Leads

For about forty years a rigid orthodoxy has controlled the social sciences. The dogma of environmentalism, labeled a consensus of silence by geneticist C. D. Darlington, has clouded the question of hereditary differences in human beings. Particularly is this true with regard to any intelligence differences. A recent *Science* review of books by Jensen, Eysenck, and Herrnstein closed with the caution that the "present social climate" must be considered in discussing racial differences. This raises the question of how *scientific* social science is. Or can be.

Herrnstein's controversial *Atlantic* article on inherited intelligence was actually a restatement of an earlier hypothesis published in 1967 in *American Sociological Review,* said to have been the first paper discussing genetic factors in intelligence ever to be published in that journal.

Kant and Descartes typified the rationalist school; Locke and others like him were empiricists. The empiricists evolved over the years, although rapidly with J. B. Watson, into stimulus-response "behaviorists." All they knew, or needed to know, was how animals (including man) responded to hunger, electric shock, reward, and so on. These basic responses could be made to account for everything, including learning and intelligence. More recently the "cognitive" school, which gives some weight to innate factors in behavior, has taken hold. However, even these enlightened specialists are so fearful of racism they hesitate to give much credit to genetic differences.

Among those attacking environmentalists are "nativist" anthropologists like Lionel Tiger and Robin Fox who believe that since most human cultures practice similar customs, taboos, and so on, much of our individual and social behavior is not chance, stimulus response, or free choice, but genetic. And language scholar Noam Chomsky has advanced a theory of language based on innate faculties. He believes that all people possess certain hereditary logical rules of communication including "deep grammar" and even thought. Thinking would thus be something we do not learn, but are born with.

Just as some geneticists believe that mutations come from within the gene and not from without, more and more scientists are beginning to believe that human abilities come not just from environment, but to some extent from within, handed down genetically over all the generations. For example, in an article titled "From the Gene to Behavior" Caltech's Seymour Benzer says: "It is well established that the genes speak strongly in determining anatomical and biochemical features. It should not be surprising if, to a large degree, the genes also determine behavioral temperament, although, of course, environmental influences can also play a role."

In general, hereditarians believe that something like 75 to 80 percent of intelligence is attributable to heredity, the rest to environment: four parts nature to one of nurture. Even more important is the basic tenet that intelligence, like height or eye color, is inherited from parents (in particular) and dependent generally on the gene pool of the relevant population.

Supposing that the heredity theory of intelligence were generally accepted, what could then be done about the intelligence of the population? Euthenics, the method of doing all we can environmentally, is still a logical approach, and this is most of what has been done consciously and with design. Eugenics, the exploiting of the mechanism of genetics, has been implemented only in negative fashion, as with immigration quotas of earlier times and the sterilization of those judged unfit to bear children. Although to many this seems a hard-hearted way to achieve success, there are arguments in favor of some control of breeding. Justice Oliver Wendell Holmes pointed out in approving California's pioneering sterilization laws, "Three generations of imbeciles are enough!"

E. W. and S. C. Reed in a paper, "Mental Retardation: A Family Study," state that 20 percent of mental retardation could be eliminated in a generation by seeing to it that no mentally retarded persons reproduced. Several generations of such a program would seem to reduce materially the number of retarded in the population. However, it would take more than ten generations, or in excess of 200 years, to remove 80 percent of genetic retards. And these are only 50 percent of the entire retarded group. Thus there is little prospect for such a eugenic program being attempted.

There have long been fears that just the opposite is in fact taking place. In 1940, Raymond B. Cattell, a pioneer developer of intelligence tests, voiced the fear that since lower-IQ individuals tend to have larger families, in time the general IQ would decline. Cattell saw evidence of such a drop of about 1 IQ point per decade in England, and 1.5 points in America. Cyril Burt echoed this fear in 1946 and predicted a drop of from 1.3 to 2.5 IQ points per generation. Contrary to these hypotheses, however, it was reported in 1942 that the intelligence of children in Honolulu *increased* nearly 20 points between 1924 and 1938! In 1932 and again in 1947 IQ tests were given to some 88 percent of eleven-year-old children in Scotland. In 1947 the IQ test showed 2.2 points *higher* than in 1932. And when Cattell himself more accurately tested his hypothesis in 1950 he found a *gain* in a thirteen-year period of 1.28 points. Another study in 1948 showed that servicemen jumped from an average score of 62 in World War I to 104 in World War II!

. . . The value of the idea of population is that it expresses the most direct hereditary and reproductive relationships among individuals and that it distinguishes a kind of evolutionary unit. It is the visible manifestation of the "gene pool." By the broadest definition, a population is a group of individuals of one species so situated that interbreeding can take place among them and which thus has continuity through successive generations.

Theodore H. Eaton, Jr., *Evolution*

5   Population Genetics

Cell genetics on a scientific basis had to wait until recent years for its flowering. Population genetics is a much older science, dating back at least to the turn of the century when Hugo De Vries (who postulated large mutation as the mechanism of evolution) and others rediscovered Mendel's work with plants leading to the theory of genes.

In 1908 British mathematician G. H. Hardy and German geneticist Wilhelm Weinberg independently discovered the mathematical relationship between the genes in a population and the genetic types of individuals, or "genotypes," that can arise by random combinations of these genes. An important fact about population genetics is that these gene population relationships can be accomplished in a single generation; geneticists do not need to know the past history of a population. It was the interplay of the newly discovered Mendelian genetics and evolution research that led to the science of population genetics as distinct from "classical genetics," which studies the offspring of one set of parents and focuses on inheritance at the family level.

A "Mendelian population" is a group of individuals who interbreed among their own members according to a certain mating system. A population possesses a "gene pool" or total available store of genes from which come the genotypes that characterize the offspring of the population. In a stable situation, the gene pool remains the same, although each generation is reshuffled into a variety of genotypes through combinations and recombinations of the genes. In the simplest form of population genetics, completely random mating or mixing of genes takes place. This is a theoretical system, however, and in real life there is much selection, rather than random mating. The other end of the population spectrum is total inbreeding.

If the gene pool of the original population is added to by migration from another population, further genetic variations will come about in the progeny of the new combined population. An obviously important factor in population genetics is the size of the breeding population. In a very large population the gene pool is stabilized and gene frequencies stay close to equilibrium. Any further genetic changes will come about only because of environmental change.

In a very small, isolated population, inbreeding theoretically leads to practically no gene changes at all in new individuals. Such "invariant" genetic types face possible extinction because of their inability to adapt to new conditions should they arise.

It is in the population of intermediate size that all genetic factors, random and systematic, are put into play. Evolutionary change is more likely to occur in this situation, which is the one we have in the world today; although the human population is large, it is subdivided into many groups with some migration among them. Genetic selection here plays more of a part than in either the very large or the very small population.

We saw in an earlier chapter that even a few gene variations could produce more different genotypes than there are particles in the universe. In a large, random mating population, we might expect great differences from one individual to the next. Indeed, if it were not for a number of forms of genetic selection, there would be far more differences among humans than now exist.

Of great importance is natural selection, the "survival of the fittest" to which Darwin and his followers primarily ascribed evolution. There are not 3.5 billion greatly different human beings probably because all these varieties could not all exist in the environment we have. Size is an example. While occasionally a giant is born, such genotypes find life difficult for environmental reasons, including gravity. Some people are born with only one kidney, but those with two have a better chance of survival and thus reproduction.

To natural factors, and to the mathematics of population genetics, humans themselves have added further constraints. Sexual selection has been carried to extremes including celibacy and homosexuality. Surgery and medicine also upset the genetic balance. So do modern transportation and the rapid mixing of formerly isolated populations.

In addition to the genotypic variety caused by combinations of genes, another kind of change comes about through "mutation" or change in the genes themselves. Historically mutation has been vaguely described as a "spontaneous" event, caused by natural radiation, accident, or unknown factors. However, man himself is becoming more an agent in mutation through artificially produced radiation, chemical mutagens, and the like. Radioactivity, air and water pollution, deleterious chemicals in food and drink, drug addiction, and smoking all may be factors in genetic mutations.

We have talked of the "random shuffling" of genes, as though this is nature's plan, since anything else would be goal-directed and thus outside the province of chance. However, it is obvious that even among lesser species than man mating takes place not on a random basis but on a fairly selective one. "Sexual selection" came in for much discussion by Darwin. Obvious examples are the plumage, vocal displays, scents, and other attractants of mating birds, animals, and fish. Of the sex life of bacteria we are not sure, but there may be some selection criteria that affect future generations.

Urbanization is a factor in population genetics. So is the mating of men and women of well-matched intelligence, and the selective mating of races. The question of abortion on a permissive basis is interesting from the genetic aspect also. Would permissive abortion gradually result in the production of a higher percentage of women who want children and a family life? Will the "zero population growth" drive also be a self-defeating exercise, eliminating those opposed to childbearing and thus defaulting to those who want to produce children?

Population genetics cuts both ways. Used in one direction it predicts the number of different types a population can produce in a generation, given a certain gene pool. It is also possible to work in the other direction, from individuals backward, to identify the mechanisms of inheritance. Human blood types A, B, and O are examples of such genetic mechanisms. These hereditary differences were discovered in about 1900, and by 1920 geneticists had identified the associated genes.

Simplifying assumptions often had to be made by early population geneticists, or the combination of many genes and many people would have reached such astronomical quantities as to defy analysis. However, electronic computers have made more accurate statistical work possible.

Stacking the Genetic Deck

As geneticists point out, in a randomly mating society the genes are "reassorted by the Mendelian shuffle" each generation. The 52 cards in a bridge deck produce a fantastic number of different possible hands, but the same 52 cards are reshuffled each deal and the odds remain the same against one player drawing four aces, or a perfect hand. This of course assumes pure random shuffling, with no help from the players. A card shark manipulates the pasteboards in his own favor; "inbreeding" of individuals does violence to the Mendelian shuffle. Centuries before anybody heard of Mendel or his factors, society had already set up taboos as to who could marry whom.

Darwin was much concerned with the results of inbreeding, which in general were not good. First, this genetic short circuit could nullify the benefits of bisexual reproduction over binary fission, the asexual splitting of a cell to form a new one. This simple duplication leads to the same individual over and over again. As the geneticist describes it, "inbreeding results in a decrease in genic variability in the inbred strain."

There are other results than merely a boring repetition of human types. Inbreeding, in addition to decreasing genic variability, also increases "homozygosity," the production of chromosomal sites at which gene pairs are identical. This doubling up results in recessive genes coming into play. A large gene pool works against the mating of a man and woman each with the same harmful gene at a certain site; in the inbreeding situation, this misfortune happens with increasing frequency, and ultimately with practically every mating. The plant geneticist knows this condition as "decreased vigor" and must take steps to eliminate it. Human society must also consider the results of inbreeding.

The Amish sect of Pennsylvania Dutch descended from about 200 immigrants who settled there more than 200 years ago. Although some Amish leave, virtually no new blood has been introduced into the sect. For such a group to survive it must inbreed. The Amish have more than survived, and now number 44,000. Certain consequences of the inevitable inbreeding are evident, however.

In 1744 Samuel King arrived in this country. He or his wife, geneticists don't know who, had one chromosome marred by a defective gene causing dwarfism. Since the gene is recessive, none of the King children showed any sign of dwarfism. Nor did their children's children. However, in this very restricted gene pool, descendants began to marry second and third cousins. Eventually a man who carried the harmful gene married a cousin who also carried it. Their unfortunate offspring inherited a double dose of the defective gene, and in 1860 the first Amish dwarf was born.

Besides being short in stature, these variations all have six fingers on each hand. Sometimes there is a sixth toe on one foot or both. Many dwarf babies also have an abnormal heart with only three chambers instead of four, plus a chest cartilage deficiency around the windpipe. One-fourth of dwarf children die of such defects within two weeks of birth. Another fourth have less severe heart defects and survive. Half appear to have no heart defects at all and may achieve a nearly normal life span. When a study was made in 1963, there were eight adult dwarfs. The oldest was fifty-eight; there were also sixteen children and teenagers. None showed mental retardation or IQ loss.

The Dunkers, a small religious sect originally settling in Pennsylvania, would also seem to qualify as a race apart because of their near-isolation from the outside gene pool. At one time they were shut off completely, although more recently perhaps as much as 12 percent of the gene pool comes in from marriage with converts or others outside the Dunker sect. This small population has been described as distinctive in many genetic characteristics, including differences of the hands, and in ear lobe attachment.

Natural Selection at Work

Once it was thought that characteristics were acquired by some kind of Lamarckian feedback from environment. More recent thinking is that the mixing of genes from the available pool, plus the workings of natural selection in a variety of forms, is the culprit—or benefactor—in the case of human differences. Giraffes obviously developed long necks not as a matter of will—although any neck-bearing creature can probably stretch it to some extent—but because those giraffes carrying genes that gave them slightly longer or more stretchable necks did better in competing for the high leaves on tall trees.

The "Heike" crab found in Asian waters, which has a remarkable likeness of a fierce warlord on its back, was cited long ago by Julian Huxley as an example of natural selection at work. (Crabs on which such markings developed frightened fishermen away!) Even Lamarck, who believed his giraffes increased the neck length of their progeny by stretching their own, would probably not have attributed crabs with the ability to cause a frightening caricature to appear on their shells. For those who still have difficulty accepting the Heike example (the writer included), a more credible example of natural selection has been observed in Great Britain and Europe. Industrial smoke darkening the trees that are the habitat of certain moths supposedly has led to a new type, appropriately named *Carbonaria*. The selection mechanism is as follows. In every population of moths there is a natural range of variation, allowed by the "play" in the genes. As trees darkened, moths that were darker survived better (lighter colors led to their possessors being gobbled up by predators) and in only a few years the dark moths predominated.

Another example is the selection of strains of insects, in America and elsewhere, more resistant to DDT and other powerful pesticides. It is ironic that even as we are warned that pesticides are wiping out some species, other affected creatures simply change their spots and come back stronger than ever.

Human genetic change is not as quickly evident, of course. Flies have a life span of only a few days, and genetic changes in them are quickly detected. A human generation lives for decades, and it requires patient historians to note any changes. Of course, changes have taken place and are taking place. It is important to properly characterize these, however. The average life span of man has increased greatly since earlier times. Anthropologists feel that we can reproduce at such an early age because there was a time when the average life span was little longer than the late teen years. Now many can expect to live more than seventy years. This dramatic increase, which has tremendous effects on mankind, is not the result of genetic mutation, or evolution, however. It indicates only that normal gene inheritance includes the potential to live longer in proper environmental conditions.

In early times man was barely able to maintain his numbers on earth. Today one of our major problems is holding those numbers down. The population explosion should not be taken as indication that modern parents are any "sexier" than were primitive peoples, indeed the opposite seems to be true, for old-timers produced far more children per family than we do. Again, the difference is in an environment that reduces the death rate and extends the life span.

Remarkable physical changes have occurred in some countries over a very short period of time. For example, in the years following World War II Japanese children gained several inches in height and added several pounds of weight on the average. This was not genetic change, however. Again the difference was nurture, not nature, except that nature endowed the once small people with the genetic potential to grow larger with better food and living conditions.

Sickle-cell anemia is a classic example of the workings of population genetics. This disease, which affects Negroes almost exclusively, gets its name from the characteristic curved shape of red blood cells in individuals suffering from it. The sickle-cell gene is recessive; one such gene from a parent is "masked" or covered up by a normal gene from the other parent. Many scientists, perhaps beginning with J. B. S. Haldane, have suggested that a single gene of this type conveys protection against diseases abounding in hot humid climates—malaria, for example.

Here is a hard choice for the genetic mechanism: needed protection against malaria is bought at the risk of anemia. However, with a sufficiently large gene pool and safeguards against inbreeding, few individuals inherit this genetic double trouble. Interestingly, some studies suggest that the disease in American Negroes has been slowly disappearing since they first came to this country, and in about the ratio that genetic theory predicts.

The Human Races

Plato described the "inherent inequality" of men. Even in recent times Count J. A. Gobineau, a cultured and intelligent Frenchman, could write his book *Essay on the Inequality of Races.* Some of the most ardent eugenicists have also been adherents of the myth of Nordic or white superiority. Race, unfortunately, has been perverted by some into racism. It is obvious that there are groups of people who differ in various ways from other groups: in color, stature, facial appearance, hair, and other outward features. More recently we have found inner differences as well. Misreading these differences, some anthropologists saw fit to set up a hierarchy of races, and even to classify some as subhuman and deserving of being killed by superior peoples.

Linnaeus, one of the first classifiers, divided the human species into four types: *americanus, europaeus, asiaticus,* and *afer* (African). Here was a start at least as scientific as the Hippocratic classification of man according to his "humors": sanguine, phleg-

matic, choleric, and bilious. The German naturalist Johann Blumenbach in the late eighteenth century founded anthropology, and with
it a more scientific division of humanity into races, a term which
Buffon had used earlier. Blumenbach saw these as races: Caucasian
or white, Mongolian or yellow, Ethiopian or black, American or
red, and Malayan or brown, increasing Linnaeus's four races to
five. By 1889, anthropologist Joseph Deniker suggested twenty-
nine races. In 1933 von Eickstedt identified three basic races, plus
eighteen subraces and three collateral races, eleven collateral subraces, and three intermediate forms.

Ernst Mayr in *Populations, Species, and Evolution* refers to
two highly competent recent treatises on the races of man, one
recognizing six races and the other thirty. Both classifications, he
says, are legitimate. Yet even the division into thirty races is by no
means exhaustive. Racial difference has been described as a species
strategy directly opposed to that of putting all our eggs in one
basket. Races differ in numbers as well as in other ways. There are
more than a billion Chinese, but only about 50,000 Bushmen and
30,000 Lapps.

Four causes are generally assigned to the production of the
various races: natural selection, genetic drift, mutation, and interbreeding. The first two are probably of most importance. A race
is, or has been, part of a larger Mendelian population, and only in
rare cases will it consist of a gene pool totally different from that
of the rest of the human world. As long as races *can* interbreed,
they belong to the same species. In the human experience, there
must have been a divergence toward more races, but now the trend
seems toward fewer. Geneticist Theodosius Dobzhansky points out
that even if we become one world genetically, there will still be
as great a diversity of hereditary endowment as before, except that
different types now found only scattered all over the world will be
working side by side. Apparently bearing out this judgment,
laboratory analyses in 1972 indicate that gene differences among

individuals from different ethnic groups (Caucasians, Negroes, and Japanese) are only slightly greater than those among individuals of the same group.

The idea persists of original, "pure" stocks of human beings which have in time become corrupted by indiscriminate interbreeding. Hitler's "Aryan race" is an example of such a myth. There are also mythical pure white races, pure Indian stocks, pure Eskimos, and so on. Notions of such racial purity are linked with the old and erroneous belief in blood lines as conveyers of heredity. We still speak of "Indian blood," "mixed blood," "full blood," and so on. Ironically, the trait for blood type is so well hidden in our genetic makeup that only a chemical test can determine it.

With all respect to well-meaning believers in racial purity, it is doubtful that the present populations of humans are deleterious mixings of a handful of original pure strains. Instead, the races described as so pure may themselves have been the results of mixing that went on for long ages. Had "racial purity" been maintained it might have led to the inbreeding weakness described earlier, producing invariant types who would have been easy marks for extinction.

The idea of races as distinct species, and even genera, was popular in older times. In fact, it was maintained that the white race descended from chimpanzees, the yellow race from orangutans, and black from gorillas! As late as the 1940s Franz Weidenreich held to a similar theory of the origin of races. In his version, the Australian, Mongoloid, African, and Eurasian races stemmed from four different lines of fossil men rather than the much more divergent apes.

According to anthropologist William W. Howells, *Homo sapiens* appeared some time between 150,000 and 50,000 years ago in the period between the last two continental glaciers. This new species began to differentiate into races at that time. Anthropological finds of "white" skulls in Europe, "yellow" skulls in China, and "Indian" skulls in America seem to definitely establish these three races as early as 35,000 years ago. Similar finds of Negro skulls have not been made, however. Human travel was occasioned by need for food, military conquest, trade, and exploration. Such mobility, in conjunction with geography and other environment, produced additional races.

We have spoken of racial intermarriage as affecting populations. There are other considerations such as the shrinking of distances by commerce and transport and communication. Race is largely geography; will it disappear as geography is increasingly less a factor in society and civilization? And what effect will there be genetically when there is *one* human race—will the gene pool have grown larger still to accommodate all the characteristics, or shrink from what it is today? Of the future of races, Howells says:

It seems obvious that we stand at the beginning of still another phase. Contact is immediate, borders are slamming shut and competition is fierce. Biological fitness in races is now hard to trace, and even reproduction is heavily controlled by medicine and by social values. The racial picture of the future will be determined less by natural selection and disease resistances than by success in government and in the adjustment of numbers. . . . What man will make of himself next is a question that lies in the province of prophets, not anthropologists.

Race does exist, far beyond a single brotherhood of man. Some writers suggest that the problem is not that there is no such thing as race, but that there are too many different races of too many different kinds. However, this condition is probably to be preferred to a monotonous sameness with all of us the same stature, hue, and disposition.

The Exploding Population

According to Marston Bates in *Our Crowded Planet:*

Throughout nature there is a balance between the reproductive rate of a given species of organism and the hazards of existence for that species. Elephants produce few young, while the spawn of oysters is innumerable. It is, to be sure, a teetering balance, so that there are often shifts in abundance from year to year or generation to generation. It is generally true, however, that biological systems have a great deal of "play" or flexibility. If the hazards of existence continue for any length of time to be greater than can be met by the reproductive rate, the species is started on the road to extinction. If reproduction exceeds mortality for any long period, the result is some sort of catastrophe— sometimes taking the form of mass suicide, as with locusts or lemmings.

There is another question about limiting population. It would seem that the larger the gene pool the more chance for experimental variety and the development of "fitter" types. The forced inbreeding of a very small, closed society seems to be harmful. What *is* the most efficient population size for the good life? For good government?

A number of scientists make the point that man is the only species seemingly unable to curb his numbers. A variety of mechanisms are said to limit plant and animal species, ranging from the standard "survival of the fittest" popularized during Darwin's time to such rampages as those of lemmings and hares. The population-control feats attributed to some bird species are difficult to believe. It has been seriously suggested that periodic voluntary flights of millions of birds are a sort of census taking that permits the creatures to count heads—and subsequently produce more or fewer of themselves as indicated! If such a mechanism does function, it would indeed put man to shame as a social engineer. But it is doubtful that animals operate any such sophisticated population-control method.

Homeostasis, an automatic balancing force somewhat like a thermostat or the float level on a tank of liquid, is generally cited as the operating force in the control of animal populations. Experiments with guppies in laboratory tanks do seem to indicate that there are factors that regulate not only the population but also the sex of surviving individuals. The mechanism here is blunt: the cannibalizing of excess members, hardly a model for human planners.

Homosexual behavior is interesting to consider with respect to population genetics. Such a tendency would seem to be self-eliminating, if it were of genetic origin. Homosexuals tend not to have as many children as heterosexuals, and this could theoretically lead in time to the elimination of people with homosexual inclinations. Yet homosexuals seem to be increasing rather than decreasing in numbers.

Some scientists, notably including Konrad Lorenz, consider competition, and even aggression, not only genetic but the basis for survival and progress, since competition generally does confer rewards for being "better." Yet if such success leads to the production of more aggressive types who in turn succeed, there might in time be all aggressives. This situation obviously does not exist, unless the genetic mechanism works toward leaders who are constantly more aggressive, with the rulers of a century ago superseded by even more aggressive leaders today.

There is stronger evidence of another mechanism; that of cooperation. The institution of the family succeeded because it was necessary for survival. A lone human could not have survived in the early days; it is doubtful that he could do so today. Man is a social animal in more ways than that cliché suggests. There may be genetic feedback operating here too, for natural selection made the family man more successful and thus more predominant.

Genetic Load: The Danger of Eclipse

There is, and has been for some time, serious concern that all is not well with the genetic mechanism—that for a variety of reasons the gene pool is being tampered with and that man is the worse for it. The term "genetic load," added to the lexicon of population genetics by Hermann J. Muller, describes the amount of damage done by deleterious genes carried in the gene pool.

Muller, who won a Nobel Prize for his work in genetics, has been described as a man "who could almost hear the sound of deleterious mutations going on all around him." He foresaw a genetic eclipse that would leave man in terrible darkness. In the dim, distant future, Muller predicted, "the then existing germ cells of what were once human beings would be a lot of hopeless utterly diverse giant monstrosities." This sad state of affairs was too far in the future to be of much general concern, but Muller argued that in a much shorter time the job of caring for the infirm would require all the energy a society could muster. Everyone would be an invalid, and the genetic load so great that it would be lethal if society had to revert to primitive conditions.

As a warning against what he saw happening, Muller formulated an "ethic of genetic duty":

Although it is a human right for people to have their infirmities cared for by every means that society can muster, they do not have the right knowingly to pass on to posterity such a load of infirmities of genetic or partly genetic origin as to cause an increase in the burden already carried by the population.

Muller also believed that leaders of society tend to have genetic traits that are least desirable in a well-ordered system. Selected on the basis of predatory rather than constructive behavior, industrialists, military leaders, and politicians were psychologically similar to the successful gangster!

Muller was not alone in his worry over the genetic load. In 1931 Alexis Carrel wrote:

. . . modern industrial civilization favors the differential decrease of the genes concerned with intelligence. It seems now to be established that both in Communist Russia and in most Capitalist countries people with higher intelligence have on the average a lower reproductive rate than the less intelligent, and that some of this difference is genetically determined. . . .

Even if we could avoid a suicidal war, Carrel feared, we would still face degeneration because of the sterility of the strongest and

most intelligent. His philosophy led to some strange pronounce-
ments, such as that the best way to increase the intelligence of
scientists would be to decrease their number.

English geneticist C. D. Darlington in *The Control of Evolution
of Man* pointed to part of the problem: "Those who were saved as
children returned to the same hospital with their children to be
saved. In consequence, each generation of a stable society will become more dependent upon medical treatment for its ability to
survive and reproduce."

Theologian Paul Ramsey is of the opinion that the gene pool is
being seriously weakened both by such environmental factors as
natural radiation, medical radiation, military radiation, nuclear
testing, power generation, and so on, and also the fact that we are
strengthening weak types, like diabetics, allowing them to reproduce. The fact that even highly antisocial behavior is not severely
punished in our society tends to downgrade the gene pool generally. People are becoming not only physically but mentally and
morally less fit. In his book *Fabricated Man* Ramsey says: "It is
shocking to learn from the heredity clinics that have been established in recent years in more than a dozen cities in the United
States how many parents will accept the grave risk of having defective children rather than remain childless."

If a husband and wife each carry a recessive harmful gene of
the same type, the chances of their having a defective child is one
in four. According to statistical probability, two of the remaining
three children would be carriers of a single gene; only one child in
four would be neither carrier nor defective. Yet, Ramsey says,
many such couples elect to have children anyway. Ramsey calls
this "genetic imprudence," with the further notation that such
imprudence is greatly immoral.

Not surprisingly, there have been a few perceptive thinkers in
each age who realized the existence of genetic "feedback" on the
part of human parents. The concept of eugenics, the conscious improvement of genetic quality of the species, is no new thing, then.

What is new is the size of populations, the amount of man's "artificial" contributions to the gene pool, and the rising awareness of the mechanics and the seriousness of the situation. In the next chapter we look more closely at this age-old dream that may now be a little nearer to coming true.

It follows from what has already been granted that the best of both sexes ought to be brought together as often as possible, and the worst as seldom as possible and that the issue of the former unions ought to be reared, and that of the latter abandoned if the flock is to attain to first-rate excellence. And these proceedings ought to be kept a secret from all but the magistrates themselves if the herd of guardians is also to be as free as possible from internal strife. . . .

We must, therefore, contrive an ingenious system of lots, I fancy, in order that those superior persons of whom I spoke may impute the manner in which couples are united to chance and not to the magistrates. . . . And those of our young men who distinguish themselves in the field and elsewhere will receive, along with other privileges and rewards, more liberal permission to associate with the women in order that under cover of this pretext the greatest number of children may be the issue of such parents.

Plato, *The Republic*

6 Eugenics

Since earliest times, society has found it necessary to establish and enforce laws to safeguard life, liberty, and the pursuit of happiness. While those conditions often seem lacking, without laws and their enforcement life might be far more chaotic than it has been and is. Down through history there have been various codes including civil law, common law, criminal law, and even "unwritten" laws that still have great force. Geneticist Hermann Muller proposed another kind of law, using the Latin term for it—*jus gentium,* the law of peoples.

Is it really up to society to police such issues as marriage and childbearing? The question seems one of degree only, for there have been laws, taboos, rules, and customs aimed in that direction since tribal times. Laws prohibiting incest surely fall under the heading of *jus gentium.* There have been sterilization laws for decades, and marriage requirements including blood tests and mental competence for longer than that. Has the time come, however, when society must begin to go much further in guaranteeing a "fit population with a healthy gene pool"? There are many who think the time for such "eugenic" measures is long past.

Heading this chapter are Plato's thoughts on eugenics. Although he did not use that word itself, he accurately prophesied the difficulty of belling the genetic cat. Indeed, it is remarkable how people have resisted the eugenic approach to human "betterment." Two generations before Plato, Herodotus had expressed the concept of "superior races." Yet, selective breeding has been rejected by every segment of society except royalty. And royalty, of course, is fading out. Francis Galton, who gave eugenics its name and articulated its cause, had no children. Countless utopian communities with selective mating rules have fallen by the wayside. Only Hitler put eugenics to widespread application with his liquidation of Jews and his encouragement of SS members to mate with "Aryan" women. Because of Hitler, and earlier racial persecutions of Jews, Huguenots, Negroes, and others, the concept of eugenics became linked with overt racism, and has not yet recovered from that damning association.

There is an appealing logic in the ideas Plato discussed long ago, nevertheless, and down through history other thinkers have returned to the sticky questions of improving the human race—or one particular race. A century ago the French historian Ernest Renan, in his *Dialogues Philosophiques* saw eugenics as a possibility: "A far-reaching application of physiology and the principle of selection might lead to the creation of a superior race whose right to govern would reside not only in its science but in the very superiority of its blood, its brain, its nervous system."

Francis Galton: Eugenicist

Renan was undoubtedly influenced by the writings of Francis Galton, who in 1869 published *Hereditary Genius* in which he suggested the conscious and planned creation of a better human race. Galton was born the same year as Gregor Mendel. This coincidence was of no help to Galton in writing his book, however, since he did not learn of Mendel's monumental genetics work until about 1900. It was the publication of *The Origin of Species* by Darwin that interested Galton in heredity, to which he devoted most of his life from that time on. A man who left his mark in a number of fields from photography to meteorology, Galton pioneered the use of fingerprints at Scotland Yard, knowing that they were a unique, hereditary stamp on each human being. In similar fashion, he proposed that heredity endowed unique "natural abilities" too.

Galton had to admit that little was known of the actual mechanics of heredity. On a purely empirical level, however, he believed in "natural ability." He also sincerely believed in racial superiority, and his writings unfortunately would become the basis for much racism in the century following. Here is a pertinent quotation from *Hereditary Genius:*

. . . The natural ability of which this book mainly treats, is such as a modern European possesses in a much greater average share than men of the lower races. There is nothing either in the history of domestic animals or in that of evolution to make us doubt that a race of sane men may be formed, who shall be as much superior mentally and morally to the modern European, as the modern European is to the lowest of the Negro races. Individual departures from this high average level in an upward direction would afford an adequate supply of a degree of ability that is exceedingly rare now, and is much wanted.

Galton was not as chauvinistic as this sounds, however. He wrote that the greatest people who ever lived were the ancient Greeks, and like Goethe he used them as an ideal to gauge other races by.

In 1883 Galton published a revision of his book, pointing out some of the errors in the first edition, including its references to Darwin's naïve "pangenesis" theory. He also defined eugenics as "the study of the agencies under social control that may improve or impair the racial qualities of future generations either physically or mentally." It is most unfortunate that he could not use the word "genetic" instead of "racial." According to Galton, eugenics was intended to give to the more suitable races or strains of blood a better chance of prevailing speedily over the less suitable. Even in his first edition he had described the potential of controlling man's inheritance:

I propose to show in this book that man's natural abilities are derived by inheritance, under exactly the same limitations as are the form and physical features of the whole organic world. Consequently, as it is easy, notwithstanding those limitations, to obtain by careful selection a permanent breed of dogs or horses gifted with peculiar powers of running, or of doing anything else, so it would be quite practicable to produce a highly-gifted race of men by judicious marriages during several consecutive generations. I shall show that social agencies of an ordinary character, whose influences are little suspected, are at this moment working towards the degradation of human nature, and that others are working towards its improvement. I conclude that each generation has enormous power over the natural gifts of those that follow, and maintain that it is a duty we owe to humanity to investigate the range of that power, and to exercise it in a way that, without being unwise towards ourselves, shall be most advantageous to future inhabitants of the earth.

Galton was knighted for his work in 1909. He believed to the end that eugenics could benefit the world, and when he died his will provided the money for a eugenics research laboratory. The man who would next take up the torch as earnestly as Galton was at that time in his second year at New York's Columbia University, and about to begin genetic research.

Hermann Muller: Sperm Banker

Hermann J. Muller was born in New York City and educated at Columbia, where he worked with Thomas Hunt Morgan on the genetics of the fruit fly. Muller found that he could increase mutations in these insects by heat, and in 1926 he added the use of X-rays for this purpose. Russia was very active in genetics, and in 1933 Muller accepted the post of senior geneticist at the Institute of Genetics in Moscow. However, he remained there only until 1937, at which time he broke with the fallacious genetic theories of T. D. Lysenko and returned to the United States.

In 1945 Muller became professor of zoology at Indiana University, and the following year he was awarded the Nobel Prize for his work with X-ray mutations. Well aware of the deadly effects of radiation on the genes, Muller began to speak out strongly against radiation exposure to X-rays in medicine and industry. After the development and use of atomic bombs his was among the loudest voices warning of the increasing "genetic load" because of radioactive fallout. In 1955 he joined with other scientists, including Albert Einstein, in pleading for the outlawing of nuclear bombs. Because of his concern over the increasing genetic load, Muller argued until his death in 1967 for some form of eugenics to balance this situation. Quite correctly he has been described as a "latter-day Galton." A major difference, however, was Muller's knowledge of the actual hereditary mechanism and of those mutagens which harm it.

Muller conceded that "genetic surgery" had a place in improving the human race, but he felt that eugenics, or "germinal selection," was the more effective procedure. In fact, toward the end of his life he was increasingly dubious of the efficacy of genetic engineering and suggested that rather than this approach we might do better to make robots. He was not alone in this belief, of course. For example, Dr. Salvador Luria of MIT has said, "I am convinced that the chances of improving human heredity by genetic surgery are much smaller than the chances of improving it by germinal choice."

While Muller is generally credited with the "sperm bank" concept, others were there ahead of him. Russian geneticist Cebowski, for example, in 1929 advocated much the same approach to eugenics. In 1935 Herbert Brewer independently thought out the germinal selection method, calling it "eutelegenesis." Sir Julian Huxley termed the technique "preadoption," and Aldous Huxley described it with approval in his novel *Island,* an idealistic antithesis of *Brave New World.* Nevertheless, it was Muller who did most of the drumbeating for VCOGP, or voluntary choice of germ plasma.

According to Muller, AID (artificial insemination donor) fell far short of the mark. Sperm donors were screened for transmissible disease, but there was no attempt to make use of the genetic potential in superior human beings. Muller felt that such donations should be much more carefully screened than the blood donations nearly anyone could make. Rather than routine specimens of germinal material from males interested only in the $10 to $25 it would pay, Muller urged banks of frozen sperm from the "great" men of the world. One of his early candidates was Lenin, a choice for which Muller was later roundly criticized by opponents of the eugenics approach. In response he suggested a seasoning period of two or more decades before any "genius seed" was used.

Genetic Eclipse?

There does not seem to be substantial evidence that Muller's "genetic eclipse," or even an appreciable waning of our hereditary star, may be taking place. However, there are many conflicting theories and reports of what is happening, and why. One writer suggests that there are problems because young people are more mobile than they used to be, and no longer marry so often within related or neighboring family stocks. Some worry that the genetically inferior produce more children and are thus slowly dominating the population. Others claim that the deleterious watering down actually began when man shifted from a hunting to an agricultural society.

Despite the confusion of theories, civilization has undoubtedly had an impact on the human race. Whereas once the balance of deleterious genes was probably held in equilibrium by natural forces, man is "unnaturally" upsetting that balance. Some scientists and philosophers think that the modern environment of public health, chemical and radiation therapy, birth control, and other factors is a monkey wrench in the gears of natural selection.

According to geneticist Leroy Augenstein, it was about 1910 when civilization "turned the corner" in medical science to the extent that a hospital patient's chances of coming out better than he went in were greater than 50 percent. This, Augenstein said, was the beginning of the end for the old Darwinian survival of the fittest.

In his book *Fabricated Man,* Paul Ramsey refers to a "genetic early warning system," and Linus Pauling once suggested that a tattoo be placed on the forehead of every young person carrying the sickle-cell gene or other deleterious recessive gene. Two young people carrying the same seriously defective gene would recognize this situation at first sight and would supposedly refrain from falling in love.

Leroy Augenstein suggests three choices:

1. Continue present medical practice, knowing that we add more and more genetic load to the population and that within five to ten generations one in ten children will be seriously defective.

2. Continue current medical practice for most people, but partially reinstate natural selection by giving no medical care to some defective youngsters and performing indicated sterilizations or abortions.

3. With all the hazards involved we can embark upon the road of genetic manipulation.

According to Augenstein, only about 5 percent of us would vote for choice number one and nearly no one would vote for number two. Overwhelmingly, the majority would be for genetic manipulation, or eugenics. Of course no such Gallup poll has been made.

Ramsey describes two types of genetic control. The first he calls "negative" eugenics, the *breeding out* of harmful genes. The second is "progressive" eugenics, the *breeding in* of good genes. Ramsey believes that genetic control should be exercised only in the interests of the husband, the wife, or the child they are having, and not for the species generally or toward control of evolution. Nevertheless he feels that "Aldous Huxley's fertilizing and decanting rooms in the Central London Hatchery [*Brave New World*] could become a possibility within the next 15 to 20 years. I have no doubt they will become actualities at least as a minor practice in society."

"Negative" Eugenics

At the International Congress of Eugenics held in New York in 1932, one of the speakers stated:

There is no question that a sterilization law enforced throughout the United States would result in less than 100 years in eliminating at least 90 percent of crime, insanity, feeble-mindedness, moronism, and abnormal sexuality, not to mention the other forms of defectiveness and degeneracy. Thus, within a century, our asylums, prisons and state hospitals would be largely emptied of their present victims of human woe and misery.

It is interesting to compare this utopian-sounding prediction with calculations made by Theodosius Dobzhansky in his book (with L. C. Dunn) *Heredity, Race, and Society*. Starting with a genetic abnormality affecting 25 percent of the population, sterilization in one generation would drop this to 7 percent; two generations to 6 percent; three generations to 4 percent; four to 3 percent. Within ten generations it would be 0.8 percent, twenty generations, 0.2 percent, and in thirty generations only 0.1 percent. If realizable, such effectiveness would seem to argue for negative eugenics measures.

There are difficulties not seen at first glance. Negative eugenics *has* been practiced for some time in two major ways: the immigration quota acts and those at the state level permitting and even requiring sterilization of those judged unfit to have children. Some critics have long felt that eugenics arguments are based on inadequate or even erroneous information and errors compounded by racial prejudices. A recent example lies in the XYY sex chromosome abnormality at first thought to be associated with criminal tendencies. Normally a male is born with an XY pair of sex chromosomes, but in rare cases an individual receives an extra chromosome from his male parent. (There are also XXY genotypes and even XXYY.) The XYY variant was soon being referred to as "supermaleness," and abnormal behavior characteristics were said to be associated with it. Indeed, possession of the XYY chromosome pattern was being argued in court as genetic insanity, as a defense against murder charges.

A paper by Saleem A. Shah of the National Institute for Mental Health, "Recent Developments in Human Genetics and Their Implications for Problems of Social Deviants," cites 200 items published in the scientific and professional literature since 1961, when the first report on the XYY chromosomal abnormality was published. The first XYY male described in scientific literature was of average intelligence without any apparent physical abnormalities and had no criminal record.

Multiple murderer Richard Speck was reported in the news media as a "supermale" with the XYY, yet subsequent reports stated that Speck was *not* an XYY male. Although several XYY males have been documented as having gross mental retardation and low intelligence, there have also been reports of XYY males with normal and even superior intelligence.

The National Institute for Mental Health issued the following statement at a 1970 conference: "The demonstration of the XYY karyotype in an individual does not in our present state of knowledge permit any definite conclusions to be drawn about the presence of a mental defect or disease in an individual. A great deal of further scientific evidence is needed." In light of all this, Caltech's Robert Sinsheimer asks whether or not the genetic counselor finding an XYY chromosomal aberration is bound ethically or morally to inform the parents of this condition.

However, there are less controversial areas where negative eugenics might have value. Paul Ramsey notes that very young women, no less than women who bear children past the age of thirty-five, have a greater incidence of chromosomal abnormalities in their children. Studies made in Japan—where the average upper age of childbearing has rapidly gone down to thirty-five, while the age when women have their first child has risen in recent decades to twenty—have shown not only a decrease in mongolism, which rises exponentially when a woman is of older age, but also a marked decrease in congenital defects attributable to childbearing at an early age.

Ramsey points out that it is insufficient merely to admit that there are hazards in genetic manipulation. Instead, we need to know that there are *no* hazards before we start doing it. This is a difficult restriction because it may take twenty years to find out the results of genetic manipulation. Rather than attempt gene therapy for diabetes carriers and sufferers after the fact, he says a better treatment would be "continence, or not getting married, or using three contraceptives at the same time, or voluntary sterilization."

The "Superrace"

As pointed out by geneticist Amram Scheinfeld, back of all positive eugenics proposals hovers the thought of producing a race of superior people, or perhaps the breeding of human beings for specific purposes as we now breed animals. This could probably be done—once we have determined the exact genes responsible for the variety of human traits—but it would demand control by the state of all human reproduction: who mates with whom, when, how often, and which children survive. Scheinfeld points out that this virtually happened in Sparta, and came close to happening in Nazi Germany.

It is a tragedy of Galton's heritage that the man who took his ideas most seriously should pervert the concepts of eugenics. Here is Hitler's philosophy in *Mein Kampf:*

> Just as little as nature desires a mating between weaker individuals and stronger ones, far less she desires the mating of a higher race with a lower one. As in this case her entire work of higher breeding, which has perhaps taken hundreds of thousands of years, would tumble at one blow.
> Historical experience offers countless proofs of this. It shows with terrible clarity that with any mixing of the blood of the Aryan with lower races the result was the end of the culture barrier.

Along with the Jewish people, Hitler also set back the concept of eugenic improvement of human beings. Yet the dream—or threat—persists. Augenstein asks: If a parent had the option of having a child with an IQ of 200, would he take it? Particularly would the parent do so if he knew that the Russians were already breeding a whole generation of 200 IQ's? This is the line of reasoning taken by Dr. James Bonner of Caltech:

> Now it's clear that if any country on the face of the earth starts a large-scale program of selecting better people, all the other countries of the world will have to get into the act too, and join up, or else they'll soon be composed of obsolete, old-fashioned people like us instead of the new super people. And so the idea will spread.

The "genetic imperative" is not that simple, however. For all its promise it is as easy to argue against as it is to advocate.

The Mirage of Eugenics

Making the best use of our talents is an appealing idea, and this concept is supposedly at the root of eugenics approaches. For example, Robert Francouer has asked the question of whether priests and nuns should be allowed to contribute sperm and ova for germinal banks, although they have traditionally been outside the gene pool. The benefit of such eugenic details is completely overshadowed by more basic genetic questions of the concept, however. For instance: two geniuses marrying will produce genius children, and if they in turn marry geniuses, the next generation will be even brighter. Two midgets will produce midgets. Two giants will have giant children. Two idiots will produce more idiots, and gradually the IQ will decline to zero. True? Not really.

Although a broad study of geniuses showed that geniuses tended to produce children above the normal IQ, there is no evidence that continued selective breeding will produce a superrace, for Hitler or anyone else. While dull mental types generally produce their kind, the offspring average several IQ points higher and may include some geniuses. Swift runners may have children who never win any races; slowpokes may sire those very fleet of foot. This is part of the miracle of the genetic system, the potential variation that can lead to evolutionary change.

To Hermann Muller's proposal for sperm banks Paul Ramsey raises the following objections:

1. The genes that are supposed to be superior may contain injurious recessives which by artificial insemination could become widespread throughout the population instead of remaining in small proportion as they do now.

2. The validity of this proposal is not demonstrated by the present-day children of geniuses.

3. It might turn out that parents who look forward eagerly to having a Horowitz in the family will discover later that it was not so fine as

they expected, because he might have a temperament incompatible with that of a normal family.

One problem of eugenic engineering, or tinkering, as some prefer to call it, is a lack of full knowledge. Aside from all this, Ramsey argues that "no one knows whether an increase in the number of intelligent men would be a good thing unless he could guarantee a comparable increase in the number of altruists." Similarly, no one knows or can know that an increase in intelligence would be desirable in the society of the remote future.

Our moral obligations to the communities of the coming centuries are "certainly less clear than our moral obligations to the communities of the present," according to Professor Martin P. Golding of Columbia University. Obviously, the more remote these generations are from us, the less we know what to desire to create for them. As Charles Galton Darwin, grandson of the evolutionist, noted in his book *The Next Million Years,* it is difficult for us to refrain from an action today for fear that it may make one of our "fourteenth descendants" suffer for it centuries hence. It has been suggested that those who attempt to plan genetic futures for mankind are in the position of sailors trying to navigate by a landmark tied to their own ship's head.

Eugenics has never really been tested over enough generations and with enough control to give it the time and numbers needed, allowing for the statistical nature of the genetic mechanism. Furthermore, there seems little likelihood that the governments that presently prevail are going to permit or encourage experiments in breeding that will lead to human analogies of thoroughbreds, of racing dogs, and other successful breeds of animals.

As a case in point, proposals such as this have been suggested for population control within a generation:

Step 1. The country would vote for a specific rate of population increase.

Step 2. The Census Bureau works this out to be 2.2 children per each woman.

Step 3. The Public Health Agencies temporarily sterilize each young girl.

Step 4. When a girl marries she is issued a "22 deci-child certificate." After she marries, the sterilizing material would be removed so that she could have children. She can then exchange ten deci-child units for each of two children. There will be authorization for two-tenths of a child left over which she can sell to another woman or otherwise dispose of.

Even assuming that this is an advisable procedure, who has the courage, the power, and the skill to implement such a plan? The same question applies with even greater force to schemes to produce "better" people. As Dobzhansky has written: "Muller's implied assumption that there is or can be *the* ideal human genotype, which it is desirable to bestow upon everybody, is not only unappealing but almost certainly wrong—it is human diversity that acted as a leaven of creative effort in the past and will so act in the future."

The potential advent of widespread genetic screening raises new and often unanticipated ethical, psychologic and socio-medical problems for which physicians and the public may be unprepared. To focus attention on the problems of stigmatiza-tion, confidentiality, and breaches of individual rights to pri-vacy and freedom of choice in childbearing, we have proposed a set of principles for guiding the operation of genetic screen-ing programs. The main principles emphasized include the need for well-planned program objectives, involvement of the communities immediately affected by the screening, provision of equal access, adequate testing procedures, absence of com-pulsion, a well-defined procedure for obtaining informed con-sent, safeguards for protecting subjects, open access of com-munities and individuals to program policies, provision of counseling services, and understanding of the relation of screening to realizable or potential therapies, and well-formu-lated procedures for protecting the right of individual and family privacy.

Institute of Society, Ethics, and the Life Sciences, 1972

7   Genetic Counseling

In August of 1972 scientists at Columbia Presbysterian Medical Center reported apparent confirmation of the belief that manic-depression is a hereditary disease. Three years earlier investigators at Barnes Hospital in St. Louis had reported presumptive evidence that a gene on the X chromosome is associated with the emotional disorder. Pursuing that evidence, Dr. Ronald R. Fieve and col-leagues at Columbia studied nineteen manic-depressives and in the families of seven found red/green color blindness, known to result from a defective gene on the X chromosome. The remaining fam-ilies displayed the blood group pattern known as "XG superscript A," also transmitted on the X chromosome. In some cases, at least, manic-depression is probably caused by a defective gene.

The announcement was important, but hardly surprising to geneticists, since there were already hundreds of human diseases thought to be caused by hereditary defects. There are bad genes as well as good ones, hardly surprising considering the complexity of the genetic mechanism and the millions of chances for error in a single human. Although genes are to blame for so many diseases, however, there is a bright side too. For genes can help detect and indicate diseases, and possibly do something toward controlling them.

Geneticist Robert L. Sinsheimer of Caltech has said that more than 2,000 different human disorders are the consequence of genetic defects. In Sinsheimer's view, schizophrenia, diabetes, susceptibility to heart disease, and even the rate of aging are genetic diseases. Because genetic disease usually strikes early, the number of "life-years" lost is higher than the amount lost to the better-known diseases. Perhaps 25 percent of the hospital beds in the United States are taken up by sufferers of genetic diseases.

According to Dr. Robert Francouer, of the quarter of a million children born each year with major birth defects, 20 percent are specific genetic, 20 percent are environmental (lack of nutrition, lack of prenatal care, X-ray contamination, measles, and so on), and the remaining 60 percent are interactions of genetics and environment.

Francouer says that fifty years ago infectious diseases caused about 60 percent of childhood deaths, and only 2 or 3 percent could be blamed on genetic disorders. Since we have controlled many of the infectious diseases, they now account for only about 2 percent of the deaths, while genetic disorders represent about 12 percent. Thus genetic counseling is of increasing importance. According to Dr. Kurt Hirschhorn, president of the American Society of Human Genetics, every human carries from three to eight harmful recessive genes. Geneticists generally estimate that about 5 percent of all Americans need or could profit from some kind of professional genetic counseling.

Mapping the Trouble

For a long time scientists have been investigating the genetic defects that cause disease. Several hundred such defects have been mapped, mostly through tissue culture studies of skin cells from patients. The use of aborted fetuses for medical research is supported by British scientists, who recommend that such use be strictly controlled and that no commercial gain should be involved. Several research projects using fetal material are under way in the United States, but there are no national guidelines governing such work. The National Institutes of Health have established a collection of cells for use by geneticists working on the problems of genetic disease. The Institute for Medical Research in Camden, New Jersey, hopes to have 200 cell lines during the first year of the project, a pilot study for a larger effort. This is useful, of course, but it is after-the-fact information. To be of most benefit, genetic knowledge should be acquired *before the birth* of the affected individual. In recent years this remarkable technique has been developed and its use is growing rapidly.

Only in 1956 was it definitely established that the normal human complement of chromosomes is 46 and not 48. This seemingly gross error was due to the difficulty of studying such tiny entities, particularly in their cell habitat. The electronic computer is of great help in cell genetics, and a particularly encouraging technique is the simulation of human chromosomes in this manner. Coupled with a new staining technique developed in 1970, computer scanning permits more rapid and accurate analysis of chromosome banding identification, a process that is difficult and tedious when done visually. Not only can the computer identify chromosome patterns; it can also be used to simulate hundreds and thousands of banding samples that go into developing algorithms for recognition of genetic patterns. Genetic counseling depends on such recognition of chromosomal abnormalities under the microscope, and proper interpretation of genetic probabilities.

Mapping our 23 pairs of chromosomes has been compared to "searching in a cluster of mazes for an invisible animal recognizable only by its smell." Thomas Hunt Morgan and other pioneer geneticists learned much about fruit fly chromosomes, and genes, but man's genetic apparatus is far more complex and much smaller. Nevertheless, some progress has been made.

The genes, rather than the chromosomes themselves, control human structure and growth. Genes direct the production of enzymes, which do the chemical controlling of our bodies, and more than a thousand such enzymes are already cataloged. There are believed to be many thousands (or even millions) of genes incorporated in the 23 pairs of chromosomes. Since we cannot see individual genes in human cells, trying to assign responsibility for disease to certain sites far surpasses the difficulty of locating needles in haystacks. Fortunately, the chromosomes themselves provide broad hints to their links of disease, and much of the progress so far in gene mapping makes use of abnormal chromosome structures.

Geneticists have succeeded in establishing the locations of more than 800 genes, and tentatively establishing another 1,000. About 80 have been assigned to the X, or female, sex chromosome. From such mapping many diseases are known to be sex-linked, or associated with the XX or XY chromosome pair that determines an individual's sex. Hemophilia is one. So are color blindness and baldness, if these are indeed diseases. It is believed that twenty or so diseases are sex-linked.

The genetic pedigree has been neatly established (with no opportunity for the purchase of a spurious one) by the identification of chromosome "karyotypes." In a karyotype, the chromosomes are arranged and numbered in pairs from 1 through 22, plus the sex chromosome pair. The "G group" in the karyotype has two sets of chromosomes, numbers 21 and 22. Sometimes an individual receives not two 21's but three. This G "trisomy" is associated with Down's syndrome, or mongolism, the condition that accounts for about one-fourth of all mental retardation in America.

Genetic Counselor at Work

Because all of our cells have nuclei, and all nuclei carry chromosomes, chromosome analysis is a simple procedure for parents. It is also possible to determine the karyotype of an unborn child. This is done by the basically simple process of amniocentesis, the withdrawal of amniotic fluid from the sac surrounding the fetus and the culture and analysis of cells from this fluid.

For about a century physicians have resorted to the therapeutic withdrawal of amniotic fluid for women suffering the condition known as *hydramnios,* an excess of such fluid. Gradually the technique has become more sophisticated, and in 1950 a team of doctors reported puncturing the amniotic sac to measure fluid pressure, with no appreciable risk to mother or child. At about the same time another doctor studied the bile pigment content of removed amniotic fluid in an Rh-factor problem. This was the forerunner of today's genetic counseling, a technique that has gone far beyond Rh-factor investigation carried out late in pregnancy. Now it is standard practice to extract fluid as early as the fourteenth or fifteenth week of pregnancy. Already some forty hereditary diseases can be detected through genetic studies of fluid cells grown in laboratory cultures. These diseases include Down's syndrome, Hunter's disease, maple syrup urine disease, and others, many of them associated with mental disorders.

In a population of expectant mothers, there are three general levels of risk. First is a small group carrying *high risk* because of chromosome malformations in their own cells, who may have already produced defective children. The risk here of a child with Down's syndrome may be in the range of one in four or five.

The *moderate-risk* group includes women past forty, and those whose husbands carry a chromosomal aberration. After a woman reaches this age, the risk of G trisomy abnormality increases appreciably. Perhaps one in a hundred such mothers will produce a defective child.

Most women, of course, are in the normal, or *random-risk*, category. For them genetic counseling is generally not necessary, or even acceptable, since there is some small risk in amniocentesis.

Genetic counseling has been put to other uses than disease counseling. For example, a request by a Swedish woman for an abortion on the grounds that the father might have been a white man she had been intimate with, rather than her Negro husband, was settled by genetic counseling. Amniotic fluid was withdrawn at the end of her fourth month of pregnancy, the fetal cells were then grown in the laboratory and typed for match with the two men and the mother. Tests determined that the baby was not that of the white man, and abortion was unnecessary.

The Agonizing Choice

Genetic counseling has been used in thousands of cases. Indications of abnormally high risk for Down's syndrome, other mental disorders, or some equally debilitating disease give parents only a choice, not an answer. It is a difficult decision at best, for many parents cannot see their way clear to end a life, even one doomed to tragedy. There is also concern over potential abuse of genetic counseling, that it will be used increasingly as an excuse for abortions, for example. This fear may be lessened by the fact that permissive abortion seems a growing trend.

The control of genetic abnormality, giving the affected individual a nearer normal life, has been achieved in Rh-factor therapy and phenylketonuria (PKU) treatment. Gout is also a disease of hereditary nature, and it is treated with new techniques for the relief of sufferers. Similar approaches may one day correct genetic defects detected by counseling, making it possible to give a near-

normal life to one otherwise doomed to the sad choice of termi-
nated pregnancy or a life of misery—for the individual as well as
those who must care for him.

Brave New World handled the disease problem in typical
assembly-line fashion: "Tropical workers start being inoculated at
Metre 150. The embryos still have gills. We immunize the fish
against the future man's diseases." Such long-range genetic inocu-
lation is still in the future.

Tests for PKU have long been given to newborn babies. This
hereditary disease, which affects about one American in 80,000
to 100,000, can lead to serious mental retardation. Fortunately it
can be controlled after birth through diet, if recognized in time.
Many states have made PKU tests compulsory. There are an estimated 100,000 female carriers in the U.S., while up to 20,000
men actually have the disease. In 1972 Dr. Oscar D. Ratnoff of
Case Western Reserve University reported a testing method for
which he claimed a 95 percent chance of finding such carriers
compared with only 25 percent accuracy for older methods.

As many as five transfusions have been given to one fetus *before*
birth where there is an Rh-factor problem. The delicate practice
of "semidelivery" has been used with some human fetuses in attempts to correct serious defects detected in genetic counseling.
However, there are many crippling hereditary diseases which cannot, as yet, be corrected. In such cases the choice facing parents
directly, and society indirectly, is that of terminating such pregnancies or living with and letting live the sad results.

Cancer and Genetics

Early in the genetic revolution one of its new "laws" seemed to
state that information could flow only *from* the DNA gene. This
made sense, in that it preserved the integrity of the master blueprint by preventing any feedback that could alter it. In 1970
researchers at the University of Wisconsin and at Massachusetts

Institute of Technology independently discovered an enzyme with the remarkable ability to cause a reverse flow of genetic information, from RNA to DNA. Here was something as exciting—and seemingly as unlawful—as making more money from the bills in our wallet instead of relying on the U.S. mint. The seeming negation of the earlier "one-way only" rule about genetic information would have been enough to mark this discovery of "reverse transcriptase" as significant. In addition, however, there was immediate application for the discovery in cancer research.

All three of the current cancer theories involve molecular genetics, and scientists now refer to the "information of cancer" which is somehow in, or added to, our genes. In the "provirus" theory, viral RNA is copied by reverse transcriptase backward into the DNA of the cell, which then begins to manufacture copies of the virus. The "oncogene" theory states that genetic information for cancer preexists in every cell, transmitted from parent to child. However, this cancer information is normally repressed and expresses itself only when triggered by radiation, chemical carcinogens, or a virus. The "protovirus" theory was proposed by Howard Temin, one of the discoverers of the reverse transcriptase enzyme. (Temin also proposed the provirus theory in 1960.) The new theory holds that cancer arises from protoviruses, or segments of genetic information, randomly brought together.

Whichever theory or combination of theories holds true, it seems sure that cancer is a genetic disease, and that genetic therapy may offer a cure for cancer if a cure exists.

The Choice of Sex

". . . a T for the males, a circle for the females, and for those who are destined to become freemartins a question mark, black on a white ground" (*Brave New World*).

Nature in its wisdom sees to it that just about as many boy babies as girl babies are born, although there are 160 male embryos conceived for every 100 female, with about 60 percent more males than females lost before full term.

Despite this seemingly fair formula for equal numbers of each sex, many parents are still not satisfied with what they get, and some 500 methods for influencing the sex of children have been dreamed up. Greek parents were told by Aristotle to mate in the north wind for boys, and the south wind for girls. The Greeks also believed that if a woman would lie on her left side she would produce a girl; lying on the right would produce a boy. This and similar theories, probably thought up by men, always associated right with males and left with females. That male chauvinist Thomas Aquinas wrote: "Woman is misbegotten and defective, for the active force in the male seed tends to the production of a perfect likeness in the masculine sex, while production of a woman comes from a defect in the active force or from some material indisposition, or even from some external influence, such as a south wind, which is moist."

The ancients believed that sea foam caused children to be born; one myth held that a saint was conceived when a star fell into the mouth of his sleeping mother. During the Middle Ages mothers desiring boy babies were advised by local wise men to drink a concoction of wine and lion's blood, mixed by an alchemist in the proper proportions. The full moon was also supposed to have the effect of producing a boy.

Hopeful rustic Americans have hung their pants on the right side of the bed for boys, the left for girls. Superstitious counterparts in Europe go to bed with their boots on to produce boys, and some French and Italian villagers claim if the man wears a hat during intercourse it will be a boy. One eastern Australian tribe believes that baby girls are fashioned by the supernatural powers of the moon, and that boys are fashioned by wood lizards. Pueblo Indians in New Mexico believed that a pregnancy could be caused by a heavy summer shower. And of course Hiawatha was born when the Western Wind quickened his mother, Wenonah.

Early in this century, Dr. E. Rumley Dawson, Royal Society of Medicine, advanced the theory that women were responsible for the sex of their children. Eggs from the right ovary would produce boys, those from the left, girls. According to Dawson's theory, the ovaries also produced eggs on alternate months, so a mother simply had to keep an eye on the calendar to produce boys or girls. Other shots in the dark were that the sex of the child will usually be that of the older parent; that ocean tides determine sex; that sweet foods produce girls, and that bitter or sour foods produce boys. (Women must have thought this one up!)

Seldom have these sexing nostrums worked, and parents like the Eddie Cantors have had to settle for a succession of girls rather than the boy they repeatedly tried for. Others get nothing but boys. This "potluck" situation is often blamed for the population explosion, and if there were some way to predetermine sex in the unborn, we might indeed have fewer children and happier parents. Whether society would be the better for it is an unanswerable question.

From time to time chemical tests for determining the sex of an unborn child have been hailed, but mothers have generally had to rely on homelier predictions such as how the baby is carried, how active it is in the womb, and so on. Sexing chicken eggs is much easier. Nevertheless, geneticists have not been idle. As early as 1889 it was suspected that there were two types of sperm. In 1938 Lancelot Hogben pointed this out in his book *Science for the Citizen:*

> In species having an XY pair in the male, measurement of the sperm heads show that the sperms are of two different sizes. This suggests that it may be possible eventually to separate seminal fluid into portions containing predominantly one or other of the sperm; the X-bearing or Y-bearing. If this could be done, the control of the sex ratio could be experimentally realizable. Recent experiments in Moscow record success with seminal fluid of rabbits.

Success has been slow for those who would tamper with nature's male-female ratio. Russian work with sophisticated electrophoresis techniques has not proved predictive, even for rabbits. However, ten years ago, Dr. Landrum Shettles at Columbia Presbyterian Medical Center did discover how to identify the "sex" of sperm. It had long been known that some sperm carried X chromosomes and some Y (an X linked with the female's X produces a girl, a Y with the X results in a boy), but positive identification of which was X or Y was not accomplished until Shettles peered through a phase-contrast microscope at sperm slowed down by an atmosphere of carbon dioxide. To his surprise and joy, he identified small, round-headed Y carriers, and larger, oval-shaped X carriers. Another surprise was that male sperm outnumbered female about 160 to 100.

Further experiments with the androsperm (male) and gynosperm (female) cells showed that while the former were faster in the great race to the egg, the latter were more resistant to the acid environment of the vagina prior to ovulation. (Another researcher independently turned up the information that the introduction of hard water in the municipal system at Merseyside, England, was followed by the production of more boys!) Shettles therefore theorized that intercourse shortly before ovulation should produce more girls, and intercourse during or after ovulation would produce more boys. Checking statistics from practitioners of artificial insemination (who arrange the attempt for the time of ovulation), he learned that indeed the ratio was 160 males to 100 females. (Dr. Sophia Kleegman has done insemination experiments which give about an 80 percent chance of producing either a girl or a boy.)

Shettles found that men with a high sperm count tend to produce more boys than those with low count. He also learned that Orthodox Jews produce more boys than does the general population, so he studied the Talmud and found very interesting information there that suggested why this might be.

The Talmud, a collection of Jewish beliefs, customs, and laws completed about 1,500 years ago, states that when a girl baby is born "the walls are crying." Perhaps partly for that reason, the book discusses the determination of a baby's sex in great detail. A major point brought out was that sex is determined at the instant of conception, and that the timing of this instant is important. A husband desiring a boy was advised to delay his orgasm until after that of his wife. A more general injunction was that women must not have intercourse during menstruation or for a week thereafter. Dr. Shettles realized that both these points were factors in the production of more boy babies, but that the second was of more importance. Abstention until a week after menstruation would result in conception very close to the time most women ovulate, and when their secretions are most alkaline. Both the Jewish customs thus meshed elegantly with the Shettles theory.

Although Great Britain's Lord Rothschild took to television to describe Shettles's findings about the two types of sperm as "a lot of tripe," Shettles persevered and came up with methods for producing the wanted sex in a child. Frequent intercourse, to within two or three days of ovulation, an acidic douche before intercourse, avoidance of female orgasm, and shallow penetration by the male should produce female children. For boys, intercourse should be as close to ovulation as possible, an alkaline douche should be used prior to intercourse, female orgasm is desirable, as are prior abstinence and deep penetration.

The system seemed to work. Shettles reported that of twenty-two couples desiring girls, nineteen produced them. Of twenty-six wanting boys, twenty-three succeeded. He believes that conscientious following of his rules can produce the desired sex 85 to 90 percent of the time.

Shettles examined only one sperm sample containing all androsperms—from a man whose family had produced only boys for 256 years! No males have yet been found who produce only female sperm. Most fathers have the potential for producing boys or girls, according to Shettles. He himself fathered four girls and three boys (without benefit of his later-discovered methods for selecting the sex of children).

After the prediction of sex, the next goal is its *control*. In 1968 scientists at Cambridge University reported on their work toward controlling sex. They described animal experiments using such sophisticated methods as electrophoresis, sedimentation, and immunological techniques, none of which had proved particularly effective in artificial insemination. Finally the expedient of examining embryos in laboratory animals and simply disposing of the unwanted sex was adopted. This suggested the possibility of using the method with human embryos to prevent transmission of sex-linked diseases such as hemophilia, simply by not allowing carriers to produce boys. Females are not troubled by this often fatal hereditary disease, and certainly a healthy girl child would be better than a diseased male, or no child at all.

Swedish researchers, using a centrifuge to separate the different types of sperm, have achieved the birth of eleven consecutive male calves by such selection. Sedimentation is a somewhat similar technique; the top of the sample would produce boys, and the bottom, girls. Dr. E. James Leiberman of the National Institutes of Health suggests a special diaphragm to pass only the sperm of the desired sex. Others propose a pill to determine the sex of a child. Dr. Charles Birch, geneticist at Sydney University in Australia, discusses the possibility of such pills for males. They would suppress one of the two sperm types, thus providing a choice of sex in the child. Dr. Emil Witschi of Iowa State University talks of somehow differentiating men into gynosperm producers and androsperm producers. Thus a woman could marry a man who could produce boys or one who could produce girls. Getting a mixture would pose a problem, of course.

Your Baby's Sex: Now You Can Choose, written by David Rorvik with Dr. Shettles, begins with this confident promise:

This book is exuberantly dedicated to the overthrow of the so-called "50/50 club," an invention of complacent baby doctors who *erroneously* tell their patients that as a "gift of nature they have a 50% chance of begetting offspring of the desired sex and that beyond this nothing can be done for them."

The book ends with this remarkable suggestion:

. . . it is not inconceivable that man will want to emulate some of these other creatures. It has frequently been said that if men could be women and women men—even for a few minutes—many of the conflicts that have simmered between the two sexes these many centuries might fade away overnight. At any rate, it is indisputable that the sort of "sexual dimorphism" under discussion here would provide life with a whole new dimension—one that might prove highly practical in the event of wars or other catastrophes that decimate one sex or the other, in the new frontiers of outer space, and in other situations in which people are isolated from members of the opposite sex.

In the world of the future, parents may no longer worry about whether their next child will be a boy or a girl: he may be both!

It is early to tell what effect the selection of sex would have on the overall population. Should the preference for boys prevail, there could be problems of a male majority that would restructure society to some extent. A recent American study showed that if parents could decide the sex of their children there would be a surplus of more than 300,000 males a year. Among the reasons given were that mothers want a boy to be a substitute for father, and fathers want a boy so he can achieve success in the male-oriented world. Such preference is age-old and worldwide. Typical is the report by colonists during early days of the British occupation of India that one native tribe killed more than 10,000 baby girls every year.

Nature, however, seems to be winning the sex battle in spite of all this stigmatization of females. At birth more boy babies have congenital defects. Thus, even though 1,050 males are born for each 1,000 females, more females survive. In 1950 there were 1 million more women than men in the United States. In 1960,

2.5 million, and by 1970 nearly 5.5 million more! The excess of women increased more than five times, while total population increased by only one-third. Women expect to live 74.6 years, and men only 67.1. There are currently 130 older women to each 100 older men; a ratio of 150 to 100 is predicted by the year 2000. There are almost four times as many widows as widowers.

This imbalance could change drastically, of course, if parents make use of sex-determining methods for their children. Sociologist Amitai Etzioni of Columbia University fears that the demand for male children is 55 to 65 percent greater than for females, and that a selection system could lead to an overproduction of boys. This would very likely affect most aspects of social life. Noting that more men vote Democratic than women, Etzioni says an over-supply of boys might doom the Republican party! He also points out that women are "culture consumers," and that a male surplus will produce a society with the rougher features of a frontier town. Another concern is the social consequence of having few or no women who have been the oldest child, or no men who have been the youngest. If all women had an older brother and realized they were second choice, might there not be psychic scars?

The Ethics of Genetic Counseling

There are other kinds of unanswered questions concerning genetic counseling. Amniocentesis involves injecting a relatively large hypodermic needle into the amniotic sac and extracting an appreciable amount of fluid. In the process it is possible to damage the placenta or the amniotic membranes, or even to penetrate the fetus itself. What effects these things have on the developing life cannot yet be positively assessed. Although doctors using the technique feel that the risks are negligible, and that little or no harm is done in the process, others question this judgment and insist that a long period of evaluation of postoperative effects is indicated.

Perplexing questions of ethics, legality, and morals are involved.

Suits have already been filed in behalf of illegitimate children and those infected with syphilis, measles, or other hereditary diseases. Now the question is being raised about potential suits arising out of genetic counseling—or even the *failure* to use it. Robert Francouer cites the case of a woman he advised that there was a chance of her child's having a cleft palate and lip—but only one chance in twenty. In spite of these good odds, and the fact that cleft palate can be repaired with plastic surgery, the woman nevertheless proceeded with an illegal abortion.

Who makes the statistical decision as to when abortion is indicated, and are these judges and their criteria correct? Should genetic screening be made law, and if so should these laws be statewide or federal? What happens when parents are counseled that there is negligible risk, and a hopelessly defective child is born? Is there recourse to legal action, or is the physician protected?

The right of privacy is also involved. Does an individual have the right to keep secret the fact that he harbors a "bad gene," or does society have an overriding right to that knowledge? We have laws in many states prescribing compulsory tests for syphilis, and for examination to establish the nonexistence of other communicable diseases before certain actions may be taken. Does "protection of the gene pool" convey similar rights of societal protection, to the detriment of individual privacy? Finally, who will make these decisions?

The point was raised long ago that rescuing those with hereditary diseases from natural selection might work to the detriment of the population as a whole. If the weak or diseased are weeded out, society in general is stronger; if defectives are permitted to live near-normal lives, including propagation, what does this do to the gene pool?

There are several ways to deal with hereditary disease. One is to eliminate or isolate carriers upon discovery, either adults or unborn. Another is to prevent defectives from producing offspring who may be similarly defective. A third approach is to take no action at all except to treat defectives to make possible a life as nearly normal as possible. All these approaches carry with them dangers of one sort or another.

Article 2 of the Declaration of General and Special Rights of the Mentally Retarded, adopted by the assembly of the International League of Societies for the Mentally Handicapped in 1968, states: "The mentally retarded person has a right to medical care and physical restoration as will enable him to develop his abilities and potential to the fullest possible extent no matter how severe his degree of disability." Would this "right" apply to the as-yet-unborn sufferer?

In 1972, Dr. H. Bentley Glass, past president of the American Association for the Advancement of Science, was quoted as follows on genetic counseling: "Screening of adult carriers of defective genes will make it possible to warn them against or prohibit them from having offspring." Yet there are many scientists who argue that such screening may lead to the creation of "genetic lepers." Recognizing the problems are such groups as the Institute of Society, Ethics, and the Life Sciences at Hastings-on-Hudson. In mid-1972, after a year of study, a team of lawyers, ethicists, and biological scientists reported several conclusions. While favoring the concept of genetic counseling for couples on a private citizen basis, the group opposed the setting of standards of "genetic normality" for the nation or the world. Carried to extremes, such screening could turn up "bad genes" in everyone. Therefore there should be no compulsory programs of screening, and individual privacy must be safeguarded in genetic counseling, so there is no stigma attached.

There were twenty-one signers of the Institute's report, including specialists in biology, genetics, ethics, law, sociology, theology, and philosophy from sixteen medical centers and institutions and universities. Prominent among them was zoologist Ernst Mayr, noted authority on genetics and evolution.

A specific concern of the Institute was a sickle-cell-anemia screening program about to be put into effect in Massachusetts on a statewide basis for school children. This disease of the blood affects Negroes mostly, and is believed to be carried by about 50,000 in America. The fear is that disclosure of a person's carrying the "trait" may lead to discrimination against him on that basis. About 2 million U.S. blacks are thought to carry the trait, but do not have sickle-cell anemia itself. In an age of easy access to all sorts of once privileged information, data banks might soon spew out genetic information that could affect marriages, employment, insurance policies, and so on.

The sickle-cell problem is complicated by the fact that it plagues black people. While counselors feel free to point out the nature of the problem, and the method of eliminating the disease (no carrier should marry another carrier, unless willing to forego having children), few if any are ready to urge direct action in this regard. An official of the National Center for Family Planning Services, associated with the National Institutes of Health sickle-cell program, has been quoted as follows: "If I hear of any counselors who are in fact counseling against having children, I think that would, therefore, be practical genocide."

Dr. Rudolph Jackson, coordinator for the National Institutes of Health sickle-cell program, while not subscribing to the genocide notion, nevertheless shies away from counseling against not having children to avoid sickle-cell anemia. Pointing out that only a 25 percent chance of anemia in the child is "pretty good odds," Jackson told a reporter: "This may surprise you, but I'd lean heavily in the direction of telling him to go ahead and have children."

The magnitude of implementing genetic counseling on a broad basis can easily be seen in the bitter fighting going on about sickle-cell anemia. When the government belatedly earmarked millions of dollars for research on the disease there were cries of "political health crusades," and sickle cell was branded the newest member of the "disease-of-the-month-club." A noted black journalist charged that his race was being stigmatized and circumscribed anew. However, PKU and cystic fibrosis, both the subject of earlier massive campaigns and testing, are "white" diseases. Tay-Sachs disease strikes Jews. And there are even a few whites, mostly Greeks and Italians (who themselves almost exclusively suffer from Cooley's disease), who also have sickle-cell anemia.

Other battles rage around the tests used in screening for the disease, and the developments of "cures." The latter controversy has led to hostilities among doctors in different camps, and even damage to professional reputations by too-rapid adoption of seemingly promising medication. These include mainly urea and cyanate, with bone marrow transplants also offering some promise. However, at the moment the outlook is not bright, and a Harvard researcher has been quoted that a cure is far in the future. This same judgment would seem to apply to any general acceptance of the whole concept of genetic counseling!

Youth is a silly, vapid state;
Old age with fears and ills is rife;
This simple boon I beg of Fate—
A thousand years of Middle Life!

Carolyn Wells, "My Boon"

8   A Cure for Old Age?

We have made, and continue to make, great progress in eliminating or controlling many diseases, but there is one that still evades us— the ultimate disease of old age, before whose onslaught all of us must yield in the end. Yet the suggestion of gerontologists and geneticists both is that old age itself is a disease, a genetic disease perhaps amenable to treatment that will slow it down! The control of aging is surely one of the most intriguing prospects of our time, and an appreciable amount of money is being spent in this direction. Yet the public seems hardly aware that such research goes on.

The Bible tells us that Methuselah lived to be 969 years old, and that longevity beyond the dreams of most of us was commonplace in early times. Although I vividly recall a statement by a memory expert in my senior high school class to the effect that "John Rovan lived to be 173 years of age," I have never been able to verify the authenticity of this ripe achievement, and the official modern record seems to be about 130 years. Even this is about double the normal "three score and ten" most of us enjoy or endure on earth. Some go willingly when the time comes, but there are others who would rather remain, living on for as long as they could enjoy life.

In the scale of longevity, human beings stand near the top. In the United States men can look forward to sixty-seven years; women, the so-called weaker sex, about seventy-five, some eight years more than males. At the short end, flies and other insects number their days in days only. Rats and mice extend life to a few years. There seems to be a rough correlation between size and age, and elephants live to a ripe old age, as do whales, except those species we are slaughtering. So do turtles, however, and some seeds seem to live (or to retain the power of life) for thousands of years. Thus size can't always be counted on as an accurate indicator of long life.

People hit their physical peak at about the teen years, and this may be part of the reason for the generation gap between old and young. It is truly said that the best thing about the good old days was that we were younger then, and many old-timers have never forgiven a system that "wastes youth on the young."

Life is beset with all sorts of traps, including accident, disease, and war. With luck and care, we can avoid these for several decades, but in the end we succumb to old age. There is a time to live and a time to die, but apparently many are not yet ready when their time has run out. Far back in history there are accounts of man's search for immortality, for the fabled Fountain of Youth.

The prescriptions have varied. Simplest was a magic elixir taken in copious drafts to confer added years and possible immortality. Exotic fruits were another cure for old age, but with the coming of biological science, some researchers began to link sex and longevity, a less than ridiculous association. Surgeries were performed toward the goal of added years; the "monkey-gland operation" gained brief notoriety. Alas, even had it been possible to successfully reseed old men it is doubtful that this rejuvenation would have permitted them extra years to enjoy their recovered powers. For eunuchs seemed to live about as long as anyone. (They may even have enjoyed a sex life too!)

The Search for the Fountain of Youth

The antiquity of the fight against time is indicated in the 4,000-year-old "Smith Papyrus," which gives a recipe "for making an old man into a youth." Actually, this prescription was merely one to cure baldness, an urge that still supports a sizable industry. But if one can restore hair, why not dunk the entire aging body and rejuvenate it all? The fountain of youth seems to have been dreamed up in the second century A.D. by the Greek writer Pausanias. The health spa for general improvement of the body became popular in Europe, and such watering places have been famous and well patronized for centuries.

Jean de Mandeville's *Livre de Merveilles,* a treatise on longevity, was popular in medieval France. And twenty years after Columbus discovered the New World, Spaniard Ponce de León organized an expedition to seek a more efficacious fountain of youth. (What better place to look than in a *new* world?) He landed in Florida, which was pretty good aim, for the place is still a mecca of old folk trying to stay young and young ones bent on enjoying that priceless condition.

Centuries ago it was thought that association with the young could fight off the rigors of old age. King David was one who used the services of young women in this way. Francis Bacon in the seventeenth century wrote *History of Life and Death,* concerned with animal longevity and the prospects of increasing it. However, about the best that science could do was to hypothesize vaguely about an "elixir of life." This goal was finally abandoned as being about as improbable as the philosopher's stone for transmuting base metals into gold. Science gave up trying to beat death, and left the field to pranksters who built devices for country fair entertainments. These featured "grinding old men young," wherein a pretty young girl turned the crank while an old codger slid down a chute and through a tunnel, hopefully to emerge a young swain again.

About 250 years after Bacon's thoughts on aging, a French physiologist shocked the profession with his revelation of injections of testicle extract—on himself. Charles Brown-Séquard was groping for the chemical facts of life, and hoped that his injections contained a magic ingredient (we know them as hormones today) that would slow, arrest, or even reverse the aging process. For Brown-Séquard this was no routine laboratory experiment but a subjective attempt to rejuvenate his own aging body. He was seventy-two, but promptly declared he felt like a man of thirty. Perhaps there was some change in his physiology, but he died nevertheless at seventy-seven, the laughingstock of the medical profession and the press.

Another bold practitioner, Eugen Steinach, tried the interesting experiment of tying off the vas deferens, not to prevent fatherhood, but to build up an extra supply of hormone in the testicles. Today, about one million American males a year submit to vasectomy as a sterilization method. Ironically, the only other effect seems to be a slight possibility of upset to the body's immunological system. It is unlikely that any sterile males are going to claim increased longevity. Steinach did live to eighty-three, although it is improbable that his experiments contributed to his longevity.

Russian surgeon S. A. Voronoff later attempted the classic "monkey-gland" experiments, implanting chimpanzee testicles in a male patient. These pioneer interspecies transplants must have led to rejection of foreign tissue, a problem that still plagues organ transplanters today. Interestingly, it was a chimpanzee heart that was first implanted in a human not long ago.

Science and Old Age

In earlier times it was often sufficient to say simply that a person had died of "old age." Nobody lived forever, and old age was synonymous with disease. But that was before modern science. Now research institutions, with sizable government grants, are involved in serious searches for a "cure" for old age. Such a cure, if it can be developed, cannot have too high a price for those keen on putting off the grim reaper. There will be long waiting lists if longevity clinics open their doors to sufferers of "old age." Are these old-timers, and the researchers, kidding themselves—is it Ponce de León and Voronoff all over again with wild dreams and cynical exploitation of unrealizable hopes?

All this interest and solicitude for aging is in great contrast to the treatment of old folks in caveman days. In those tough times (which later must have been referred to wistfully as "good old days"), the worn out granny and the grandsire with a broken leg might be left by the side of the trail to fight off the saber-toothed tigers as best they could. Old age was a drag, and its sufferers couldn't contribute anything to the family or the clan.

Now, with surpluses of everything, including time to think about how little time we really have, man is becoming more concerned with his plight as a geriatric. By the nature of things, all who survive the earlier hazards of life join this aging band sooner or later. It may be hard for half of us to understand the problems of the other sex, or for those of one race to sympathize deeply with those whose skin color or other outward appearance is different. But *all* of us are getting older, every minute of every day. Pensions for ball players or railroad engineers command the interest of those particular segments of the population. Food for the undernourished appeals to the hungry portion. But age is the ultimate common denominator that brings us all up short. Probably few people in good health have not at one time or another, and often continually, dreamed of living on and on and on.

This is the kind of progress society has made from a time when it accepted dying as something as natural as being born, when the old had to go to make room for newcomers. Fairly recently Social Security was set up to afford some financial help to the aged; Medicaid and Medicare added health protection. But the National Institutes of Health and some other organizations are seriously seeking the causes of old age, and trying to learn if there is a cure for the condition.

Does the Human Machine Just "Wear Out"?

According to Dr. Alex Comfort, director of research in gerontology at University College, London, if we could maintain the youthful vigor possessed at our physical peak, half of us would live to be not seventy, but ten times that age. And the other half would live to be 1,400! The theory behind this startling belief is that death comes as the result of accident or disease striking down a machine no longer able to take the beating it once could. Old age is not a cause of death; it merely makes us susceptible to death or harm from other causes we earlier could have fought off.

There is an intriguing controversy between health seekers in the matter of exercise. On the one hand are those who say that the heart was created with just so many beats in it, and we had best take life easy if we don't want to strain the old pump. The other school believes that proper exercise will keep the heart—and the rest of the body—in better condition and thus lead to longer life.

The first idea is similar to that behind the warranty on a new car, guaranteeing a certain number of miles of operation. Does a human frame simply wear out after so many "man-miles," or the burning of so many calories of fuel? This latter theory has been carefully investigated. In 1908 German biologist Max Rubner published figures comparing the consumption of calories with life span, for man and various animals. Rubner was convinced that longevity was linked to metabolism, or the rate of living, a correlation that agrees with the theory that resting prolongs survival.

It has since been pointed out that there were serious discrepancies in Rubner's figures; man, cats, and dogs do not seem to fit the scale meaningfully. Still, the theory is interesting as an attempt to find a cause for aging, and worth comparing with the idea of burning the candle at both ends for a short and happy life. Radiation dosimeters are used to measure the cumulative dose one has been exposed to. Alex Comfort has suggested that analogous "life meters" might be attached to animals and men to measure energy expenditure and thus indicate roughly how much of life had been used up.

It is helpful to consider the body as a machine, although admittedly a machine like no artificial one ever made. The human heart, for example, operates with the vigor and output of a fuel pump, without a stop for as long as a hundred years or more. Recently we have begun to make repairs and replacements on the heart and other organs; such maintenance work may have an effect on longevity as it becomes more effective. Other analogies that come to mind include recharging a battery, or winding a watch. But the body is not that kind of machine. Neither is it like the wonderful "one-hoss shay" of Oliver Wendell Holmes, so carefully designed that each piece lasted as long as the others. The marvelous buggy endured for years past normal—but when it went, it went in a dramatic near-explosion of everything letting go at once. Humans often die when only one vital component fails.

Since life is a chemical process, and man is a heat engine, temperature should thus be a factor in life. This consideration has been checked out, too, to find if cold-blooded animals live longer than warm-blooded ones. Conclusive results have not been obtained, for the problem has been found to be more complicated than was supposed. While cold-blooded animals do live longer at lower temperatures than high temperatures, this longevity cannot entirely be attributed to temperature. However, a number of creatures hibernate, and it has been pointed out that some insects can remain in one stage of development for very long periods of time under certain conditions. There may be clues to longevity in these "resting" phenomena.

A strange kind of longevity has been suggested by those who advocate the freezing and cold storage of humans to preserve them until medical science can cure their presently fatal ills. Such icy slumber is hardly living, and furthermore the total metabolizing duration would probably be the same when we added up all the bits and pieces of such a patchwork of life spent half in and half out of the freezer. However, no less a scientist than Ben Franklin did say that he would very much like to be able to be put away for safekeeping and brought out periodically to see how the world was progressing!

Although the suspended animation of humans has barely been approached, it is possible to affect the growth, the rate of development, and even the longevity of invertebrates in the laboratory by raising or lowering the temperature. A variety of creatures, from the tiny rotifer to hamsters, can be put in a state of suspended animation by freezing. Dr. R. R. J. Chaffe, of the zoology department of Missouri University, conducted experiments with hamsters and found out that their habit of hibernating is under genetic control. It has been suggested that if we can isolate the gene that controls hibernation, we might investigate such activity in human beings.

Diet has great bearing on rate of growth, development, and also on longevity in animals. In the 1930's Cornell University researcher Clive M. McCay demonstrated that rats and mice could survive longer if fed less than the amount needed for normal growth. On a restricted diet they grew more slowly, developed more slowly, and lived longer. A normal rat is senile by age three, but McCay was able to keep them alive for nearly three years in a state of immaturity. When he restored normal feeding, the rats quickly grew to maturity and adult size. They then lived another three years, for a lifetime double that of normal rats. The pulse rate of the rats dropped to 340, about 100 below normal, in the near-starvation program. Rats starved *after* attaining the age of puberty showed no increase in longevity.

McCay reported that the starved rats remained young-looking and alert. Once again fed enough to permit normal growth to adulthood they did not suddenly age and die off, like the people who fled from Shangri-la in James Hilton's novel *Lost Horizon,* to die of sudden exposure to the environment outside that mystical Tibetan utopia.

More recently, rat-starving experiments have been repeated, both here and abroad. In less drastic programs, with fasting one day in three or four, growth of the animals is not arrested. Longevity is still boosted, however, although only about 20 percent rather than 100 percent with the more drastic fasting program of McCay. Drs. Frederick Hoelzel and A. J. Carlson of Chicago did these experiments. Hoelzel is said to have been so impressed by the findings that he took to dieting himself. In an analogous experiment, Max Hartmann at Tübingen, Germany, experimented with preventing an amoeba from reaching "critical" size for cell division for 130 days. Thus, it was claimed, its life span had been increased 65-fold over the normal two-day span.

It has been suggested that the arrested maturity approach be tried with humans by keeping them children until about age seventy, then feeding them well and watching them live for another lifetime. At least one potential problem is suggested by the fact that the starved rats did not develop sexually upon being properly fed. It is also questionable what sort of life perennial human teenagers would lead up to age seventy, and whether they could be useful as well as youthful members of society.

Experiments with starving humans have generally been carried out only in the prisons of totalitarian countries, where such methods have led not to increased longevity but the reverse. A good many of us in America and in other developed countries eat far more than we should. In the process we probably shorten our lives, insofar as we make ourselves more susceptible to heart trouble and other diseases. Whether we can extend the argument and say that further weight reduction will yield added benefits remains to be seen, and may be a dangerous experiment for human subjects. Involved are the problems of the exact stage in development at which starvation increases longevity, and the question of whether the starvee would lead a life so subnormal that he could not tolerate it even for the deferred reward of longer life.

Some researchers question the effect of radiation in the process of aging, and logical reasoning suggests that it may be involved. Mutations are generally harmful; radiation may cause cell body damage or even genetic damage that would accelerate aging. An obvious approach is to find the longevity of doctors involved in therapeutic radiation techniques. The first such study seemed to indicate that these practitioners did indeed have shortened lives; however, subsequent studies are generally thought to be inconclusive.

The Genetics of Aging

Francis Galton suggested that the best way to live to a ripe old age is to be born in a long-lived family. This is correct scientifically, for longevity seems to be inherited, as is body shape, hair color, color blindness, and intelligence. There is some evidence that the age of parents at the time of conception is a factor in longevity too, since older mothers are more likely to produce imperfect children. What seems desirable, then, is a set of parents destined to live a hundred years, who produce children at an early age. Experiments showed that old rotifers produced young that were shorter-lived to the point that the strain soon became extinct. On the other hand, young rotifers produced offspring that lived about 100 days, four times the normal lifetime of the rotifer.

There seems to be a correlation between sexual activity and longevity; mating rats live longer than celibates. This link is more difficult to ascertain in mankind, although it might be guessed that the same situation prevails. There are tactful ways of checking; statistics in England for the 1930s indicated that Protestant ministers had a death rate of only about 75 percent that of Roman Catholic priests.

It was once believed that single-celled creatures, barring accidents of various sorts, might live practically forever. Multicelled creatures, on the other hand, were thought to have sacrificed immortality for a more complex, specialized life. Now, however, it is suspected that there may be multicelled creatures with the gift of extreme longevity, just as there are single-celled creatures that die of "old age." The sea anemone has been observed to live as long as 180 years in the laboratory, and this specimen died only because of a water-system failure, which would have also killed a younger anemone. Alexis Carrel kept bits of chicken heart, multicelled highly specialized life, alive for decades in his test tubes.

The anemone survives because it can replace all its cells continually. In reality, it is not old at all, since it is constantly new, constantly changing. In man, however, many cells cannot replace themselves. The brain cells and those of some other organs are constantly dying off, obviously a factor in senility and ultimate death.

One theory of the mutation mechanism is that radiation destroys part of the chemical information content of the genes. Another is that *natural* decay of genetic information leads to damage and to death. Still another theory is simply that the genetic system does not include any instructions past the age of reproduction. Not that there is a built-in "self-destruct"—there is just *nothing* built in for advanced age, and we simply die for lack of a program!

Russia's Zhores A. Medvedev wrote in 1961:

It is natural that if aging is a general biological phenomenon, inherent to all forms of living matter, then it must be associated first of all with certain characteristic features of the metabolism of proteins and nucleic acids, since it is precisely the synthesis and metabolism of these compounds which constitute those processes that unite all living matter.

. . . we believe that the aging process of individual biological systems should be examined against the background of a general immortality of living matter, i.e., should be considered as the result of a genetically inbuilt weakening of the control of the exact reproduction of the specific nature of all structures during processes involving a rejuvenation of these structures. . . .

There seems to be general agreement with this evaluation, yet a surprising number of quite different theories of aging have been advanced. In fact, Dr. Alex Comfort, citing some 120 different ones, says that the study of aging has been "ruinously obscured by theory." By 1960 the National Institutes of Health were sponsoring 700 aging-related projects, at a cost of more than $16 million. This was triple the amount spent in 1954, and the figure has about tripled again in the last decade.

Swiss gerontologist Frederick Verzar has shown that collagen, the protein that constitutes about 40 percent of the human body, becomes progressively "cross-linked" with age. Dr. Johan Bjorksten has suggested that cross-linking of other body proteins occurs as well. Bjorksten therefore suggests breaking down of cross-linked molecule groups and excretion of them from the cell to make way for new molecule production. This might be accomplished with enzymes, and it has been suggested that soil bacteria might contain suitable enzymes to break down cross-links.

Gerontologist Dr. Bernard Strehler in 1971 offered a somewhat similar theory. Some cells make poisons called "chalones" which retard their ability to reproduce, according to Strehler. Animal tests have indicated that antibodies can be manufactured to purge these chalones and restore the reproductive capacity of cells. Strehler, incidentally, dismisses rat longevity studies as having no application to human beings.

National Institutes of Health officials say two genetic factors affect aging. First is the hereditary genetic makeup; second, the continually degrading genetic mechanism of body cells. Radiation could cause this degrading, as could some chemicals. Scientists don't look to mutations for many of the answers, however. While radiation is thought to shorten the life of man, 4,500 roentgens double the life expectancy of the common fruit fly. A dose of 100,000 roentgens doubles the life expectancy of the simple, hydra-like *Campanularia flexuosa*. Biologist Philip Handler believes radiation accounts for no more than 10 percent of human mutations, and thus is unlikely to be a significant component in aging.

World of Senile Citizens?

The current preoccupation with extending life seems ironic. With things already admittedly tough for the aged, why the drive to produce twice as many senior citizens, who will then produce twice as much total suffering? Recalling the Wandering Jew, it would seem that immortality might not confer the hoped-for blessings. More than a hundred years ago Arthur Hugh Clough wrote the following warning: "Thou shalt not kill, but needst not strive / Officiously to keep alive." More recently Sir George Pickering, professor of medicine at Oxford, took this view:

The goal of medicine is that of indefinite life, perhaps in the end with someone else's heart or liver, someone else's arteries, but not with someone else's brain. Should it succeed, those with senile brains and senile behavior will form an ever increasing fraction of the earth's population. I find this a terrifying prospect. We may well ask ourselves whether it is time to halt the program of research and development which will make such a thing possible. We must face up to the probable consequences of our ideas and ideals and be prepared to revise them.

Leroy Augenstein, who predicted that we will be creating new hearts and livers from our own cells within this century, suggested that it is possible to keep giving a person such replacement parts.

He then asks if such a person who eventually refuses replacements, say when he is 350 years old, would be committing suicide.

Dr. Alex Comfort doesn't see the outcome of aging research in this dim light at all. His goal is for the maintenance of human vigor, rather than the production of human vegetables kept from death long past their time. Most mammals live five or six times as long as their age of maturity. In man this could mean 100 to 125 years, and a rare handful of us have already attained that ripe vintage. All our medical advances have added no years to man's potential longevity but simply permit more men to reach this endowment. Comfort, who has been described as the "most responsible and unbiased gerontologist in the Anglo-American community," made this assessment of the aging situation in *New Scientist,* December 11, 1969.

. . . The patching-up of single age-dependent diseases is both expensive and of predictably limited use, judged by the length of further vigorous life.

At the same time, the high stability rate of vigour loss in every species investigated also suggests that the uniformity of the process is not simply a statistical derivate of underlying diversity. It suggests that there is a clock or clocks, and that by tampering with this mechanism the timing of degenerative changes could be altered, not just piecemeal, but across the board. The conviction that such a project is worthwhile has grown steadily over the past 20 years. In many countries, investment decisions are being taken, or about to be taken, about its future scale. Several means are already known by which the lifespan of rodents can be increased by 20–40 per cent, using relatively simple dietary or chemical techniques, and one (radiation) by which ageing can apparently be hastened. The exact relation of these manipulations to the rate-of-ageing clock may be debatable, but it seems highly likely that similar means would affect the human lifespan in a similar way.

What Comfort is interested in is "rate control," since longer vigorous life can probably be obtained in no other way, and because it should be easier to slow a genetic rate than to rewrite a genetic program itself. He feels that we have about reached a stone wall in the disease-control approach, and that we must switch to rate control instead. This would slow the rate of vigor loss in humans, and thus aid the human frame in holding out against the hard knocks of environment.

Comfort suggests that the reason for slow progress in increasing human longevity as compared to success with shorter-lived animals is the length of time that human experiments would take. Donors are willing to fund five- and even ten-year programs, but waiting for results from human experiments that will take twenty and thirty years is beyond their ability. He points out that tests now exist (developed to measure quickly the aging of Hiroshima survivors) that can assess rate-of-aging changes over a three- to five-year period. Here are his guesses of progress that could be made with "decent human experimentation":

By 1990 we shall know one proven way of extending vigorous life by about 20 percent.

The agents will be simple and cheap, and not require grafts, or intensive care units.

The direct application of such research would be at about the same rate as that of antibiotics worldwide.

All existing medical services and governments will elect to apply it, or will be unable to prevent its being used.

In answer to those who fear that the old are past help by rate-control techniques, Comfort encouragingly suggests that it might prove best to begin therapy at, say, age fifty, since a genetically programmed aging change (if there is one) would probably not come into effect sooner than this.

Broadly speaking, ageing probably resides in cells or long-term molecules or both. If it is a "noise" process, like the wear in a record, we need to cut the rate of noise injection—in other words, lubricate the needle. If it is programmed, like a record which, once played, cannot be re-run, we need to run the record more slowly, though not so much so as to spoil the music.

Comfort also makes the ironic concession that in light of the way cigarette hazards are tolerated because it is inconvenient to safeguard our health, we must make age enhancement available with no effort! He cites three other studies of age control, which have timetables ranging from 1993 and 2023 for extension of human life, but he himself is more optimistic. He believes that control of aging will come before the immunological or chemical control of all malignancy. Many other scientists seem to agree. For example, Nobel laureate Sir George Thomson has said, "There does not seem to be anything in the nature of the reproduction of tissue which demands its death. . . . I believe medical research should spend increasing effort on the prevention, which at first will be postponement, of old age." And in 1966 Dr. Carrol M. Williams of Harvard, active in "juvenile hormone" work with insects, stated that "the day is not too far distant when we will be able to treat senescence as we now know how to treat pneumonia." That day should now be nearly a decade nearer.

The effects of increased longevity on population problems cannot be overestimated, and might have to be balanced by a curb on, or even a cessation of, the production of new life. As Gordon Rattray Taylor noted in *The Biological Time Bomb:* "Death is an evolutionary rather than a biological necessity." If man intends to tamper with evolution—and he has indeed been doing it one way or another for some time—perhaps the necessity is not even evolutionary any longer.

On Rack 10 rows of next generation's chemical workers were being trained in the toleration of lead, caustic soda, tar, chlorine. The first of a batch of two hundred and fifty embryonic rocket-plane engineers was just passing the eleven hundred metre mark on Rack 3. A special mechanism kept their containers in constant rotation. "To improve their sense of balance," Mr. Foster explained. "Doing repairs on the outside of a rocket in mid-air is a ticklish job. We slacken off the circulation when they're right way up, so that they're half starved, and double the flow of surrogate when they're upside down. They learn to associate topsy-turvydom with well-being; in fact, they're only truly happy when they're standing on their heads."

Aldous Huxley, *Brave New World*

9   Genetic Engineering

Genetic counseling, even when held to a literal interpretation of the term, is a great human achievement. Knowledge of the probable results of mating should be useful for bringing about individual and family happiness, and at the same time perhaps a healthier gene pool for the entire population. Armed with this sort of knowledge the logical next step is from counseling in genetic matters to outright action—that of genetic "engineering."

About ten years ago the biological time bombs began to go off in a number of areas. NASA's director of Biotechnology and Human Research, Dr. Eugene B. Konecci, made this suggestion: "Our understanding of the chemistry of genes may soon enable us to modify or enlarge organs, thus accelerating evolution to a state where man could successfully survive in strange environments." Here was a plan to outpace natural evolution, to intentionally and quickly produce different kinds of human beings by genetic engineering. "Planned planethood" became a catch phrase, and Dr. Kenneth Heuer, as astronomer, wrote a book called *Men of Other Planets,* describing how life could be tailored to environment.

Dr. James T. Bonner of Caltech suggested that we would enlarge and otherwise improve the human brain genetically. Dr. Philip H. Abelson, editor of the journal *Science,* wrote matter-of-factly about the coming biological revolution in which man might "redesign" himself. Embryologists Landrum Shettles and Daniele Petrucci grew fertilized human egg cells for as long as two months in "test tubes." Geneticist Joshua Lederberg said he saw nothing fundamentally different between vaccination with a live virus and the introduction of new genetic information into an individual. Since compulsory vaccination has been justified, so should genetic alteration to alleviate problems of genetic disease.

Lederberg also remarked on the distinct probability of producing identical human beings in any number from "cuttings" of the original model. This startling prospect was quickly enlarged upon by many other scientists. At UCLA Dr. Elof Carlson predicted that even mummified bodies could provide starter cells for reconstruction of great men of the past: "As an example, we certainly will be able to reconstruct King Tutankhamen from his Egyptian mummy." And Dr. James D. Watson, who with Francis Crick unraveled the structure of the DNA molecule two decades ago, predicted to a science panel of the House of Representatives: "A human being—born of clonal reproduction—most likely will appear on the earth within the next twenty to fifty years."

Not all of those in the biological sciences accepted this remarkable genetic prospect, of course. Some pointed to a host of sizable problems remaining before the millennium could arrive. Hermann Muller was among those pessimistic about the probability of genetic engineering. Pointing out that the weakening of the gene pool by keeping alive serious mutations would lead in the end to such a collection of pitiful relics that "it would be easier and more sensible to manufacture a complete man *de novo* out of appropriately chosen raw materials," Muller favored the eugenics approach of selection of good genes in the production of children.

Geneticist Dr. T. M. Sonneborn of the University of Indiana in 1962 had concluded a paper on the "new genetics" with this pessimistic summary: "What then may we conclude as to the implications of the new genetics for the future of man? That our profound ignorance and the special problems that arise in dealing with man—even with single human cells—make the possibility of genetic engineering in man seem very far away indeed." Sonneborn felt that it probably would not be feasible to use such genetic therapy on man as had been demonstrated with bacteria, since bacteriologists "ruthlessly throw away the million failures and keep the one success," a wastage not tolerable in dealings with human beings.

Yet late in 1972, as opening speaker at an interdisciplinary conference on "Ethical Issues in Genetic Counseling and the Use of Genetic Knowledge," Sonneborn suggested that moral and ethical problems posed by the genetic revolution already sorely strained the limits of knowledge, techniques, and cultural wisdom. The possibility of genetic engineering was not far away at all; it was in fact upon us, and we were hardly prepared for the confrontation.

Dr. Sonneborn disagreed with Hermann Muller's belief that genetic surgery would be of little use because not enough was known about the mechanics of heredity. Sonneborn pointed out that while little is known about the sites or structure of the genes in pneumonia bacteria, we have nevertheless been able to genetically manipulate the organism. Such a "semiblind" scientific approach might also be developed for replacing germ-cell progenitors and lead to control of human heredity.

Science fiction long ago produced tales of "genetechnicians" whose delicate instruments excised harmful brain cells to make better humans. Such genetic surgery continues to be talked about, and now it is not just the fiction writers who describe fine-beam lasers blasting out bad genes. Other proponents of genetic engineering suggest "chemosurgery," since natural activation and repression of genes is thought a chemical process. The first chemical mutagens were discovered by researchers in the U.S.S.R. about 1937. These were only slightly mutagenic. About ten years later, other scientists in the U.S.S.R., and in Britain identified highly potent mutagens including alkylating substances and formaldehyde. Since then hundreds of mutagens have been found by numerous geneticists all over the world. All these are capable of increasing, sometimes to a great extent, the *number* of mutations arising in an organism, but they do not produce new *types* of mutation.

The fact that DNA itself has mutagenic properties was discovered by chance. Professor Serge Gershenson, head of the virology section of the Ukrainian Academy of Sciences, when studying the fruit fly about three decades ago, added calf thymus DNA to larvae with the unexpected result that it proved to be strongly mutagenic. This DNA produced visible mutations with a frequency at least 30 times greater than that occurring in untreated subjects, and their character differed entirely from mutations commonly observed. Summarizing his work in 1966, Gershenson observed that DNA from various sources should induce different mutations in different genes. RNA, certain proteins, and even synthetic polymers might also act as genetic control agents.

A year later, Dr. Bernard Dixon, a British microbiologist, described his work with protein colicines in which several enzymes were used to repair damaged DNA strands. If both strands of DNA were damaged, repair was impossible. But if only one, it was possible for the repair enzymes to remove damaged nucleotides and then create new ones to fill the gap. Much interesting research goes on, but it is more probable that special viruses will do the trick, "insinuated" into faulty genes where they will add the proper instructions to turn a defective organism into a normally functioning one.

Gene Therapy

In an age of organ and tissue transplants, it is hardly surprising that "gene transplants" are beginning to be made. The theory behind such sophisticated therapy is simple. The body is to a large extent a chemical machine, created and controlled by enzymes produced by genes. For the individual suffering from phenylketonuria (PKU), the chemical imbalance can be corrected by diet. Unfortunately, not many hereditary diseases are amenable to such simple solutions. The alternative is to "replace the bad genes with good genes," to give the sufferer new genetic blueprints that will properly control his body. Transfusing blood in an infant is no simple thing, but a highly sophisticated procedure accomplished only with great skill and care. Yet gene transplants or DNA transfusions make it seem simple by comparison.

In 1970 the first attempt at gene transfusion was made on two children suffering from the hereditary disease hyperarginemia, a deficiency of the enzyme arginase. One result of the disease is mental retardation. It had been learned in 1958 that injecting Shope papilloma virus in rabbit skin stimulated the production of arginase in the cells. Laboratory workers exposed to the Shope virus were tested subsequently for arginine (this opposite of arginase builds up in the blood and spinal fluid of hyperarginemia sufferers) and found to have lower than normal concentrations. No harmful effects were indicated, and by 1970 researchers implanted the virus in the children mentioned. Results thus far are not conclusive. In the meantime, the experiment has been criticized because Shope virus produces cancerous lesions in laboratory animals, although this condition is not known to have developed in human beings. Similar gene therapy is a possibility in treatment of cystic fibrosis, but here again the virus under consideration also carries the potential to cause usually lethal cancer in rabbits.

In 1971 another gene transplant was attempted, not directly on a human patient in this case. Researchers at the National Institutes of Health in Bethesda, Maryland, introduced genes into living human tissue taken from a sufferer of the hereditary disease galactosemia, an enzyme deficiency similar to PKU. Its victims cannot metabolize a simple sugar in milk and other dairy products, and infants known to have galactosemia must quickly be put on a milk-free diet to prevent malnutrition plus possible mental retardation and death. As with the earlier transplant, geneticists used a bacterial virus. Added to a tissue culture from a galactosemia sufferer, the "gene transplant" invaded the diseased human cells and apparently caused production of the lacking enzyme.

Gene transfusions or transplants promise exciting developments in the treatment of hereditary disease, perhaps soon. The correction of many kinds of mental retardation and other crippling diseases is a worthwhile goal, and should be attempted. However, there are dangers, as has been pointed out, including that of rushing into gene therapy too quickly and with incomplete knowledge. Only in certain cases can gene transplants be of help—for example, when a patient's own genetic equipment is producing an inactive form of the relevant enzyme. Genetic diseases may be more complex than they seem in early study. In PKU, for instance, the concentration of phenylketones sometimes drops to a safe lower level; some sufferers with a high level have normal intelligence. Furthermore, great care must be taken in selecting viral or other gene carriers that will produce no harmful or unexpected side effects.

In May, 1971, 46 scientists from the United States, France, Sweden, and Switzerland attended a conference on gene therapy at National Institutes of Health's Fogarty International Center for Advanced Study in the Health Sciences at Bethesda, Maryland. NIH scientist Dr. Ernst Freese outlined the background situation, pointing to the significant fraction of human beings who suffer from hereditary diseases, including sickle-cell anemia, diabetes, schizophrenia, hypertension, and cancer. While it is possible through genetic counseling to detect harmful traits in prospective parents, society has thus far done little toward eliminating the possibility of defective children born of such couples. Instead, treatment is given after birth by drugs, diet, and so on where this is possible.

Dr. Freese pointed out that even with the use of amniocentesis in all pregnancies, only a small proportion of hereditary defects would be discovered, since many of them show up only in certain organs, or are not expressed until later in life. Thus most hereditary diseases will probably persist, with only certain "glamorous" exceptions responding to treatment with drugs or diet. For the remainder, gene therapy appears to be the only hope.

As with organ and tissue transplants, the transfusion of healthy cells from a normal person to a sufferer of hereditary disease would be prevented by the rejection of foreign tissue by the immune reaction (except in the case of identical twins). There are two alternatives. It might be possible to take cells from the diseased person, transform them chemically or otherwise into cells expressing the normal genetic information, and grow them in large numbers in the laboratory. They could then be returned to the patient and restore his ailing body by producing the proper enzyme. The second, and more feasible, approach is to treat the patient with viruses that supply the missing genetic information. The first such virus therapy will likely be in restoring mobile, multiplying cells such as those in bone marrow.

In 1969 Leroy Augenstein suggested the obvious possibility of injecting some sort of correcting virus into a developing embryo to rectify a genetic flaw from parents. Conceding such possibilities, NIH conferees were not yet ready to accept them as proper genetic engineering. The reported consensus of conference members was to eliminate genetic defects wherever feasible by genetic counseling or abortion, but not to use gene therapy to "rescue" detrimental genes and maintain them in the gene pool. This would rule out the treatment of embryos in the womb. For now, gene therapy should be used only to alleviate genetic defects after birth. Later it might be possible to use gene therapy in the liver, and in other cells that multiply in the adult. The NIH conference set a probable timetable for such work at between one and two decades hence.

Dr. Alfred Knudson, Jr., of the University of Texas Graduate School of Biological Medicine Sciences at Houston, grouped the problems of gene therapy according to several genetic causes: chromosomal, Mendelian, polygenic, and somatic-genetic.

Most chromosomal defects are congenital, and would require treatment so early in pregnancy that gene therapy would probably not be profitable. Termination of pregnancy would be advised in such cases.

Mendelian defects are often associated with prenatal defects too and in such cases would require prenatal therapy. For those disorders that develop postnatally, gene therapy might be useful. Knudson pointed out that most enzymatic disorders can be circumvented if the infant can be provided with even 10 to 20 percent of normal function. However, in extreme cases, abortion might be preferable.

Polygenic disorders (in which a number of genetic factors contribute) including schizophrenia, hypertension, and diabetes mellitus appear to be diseases in which gene therapy can make a highly significant contribution. Somatic mutation may result in such defects as cancer, and disorders of the immune reaction system and the aging process. Knudson felt that gene therapy might be applicable here. Sufficient animal experimentation should be done so that treatment of human beings will provide benefits that clearly outweigh potential and unknown risks.

Concerning misuse of gene therapy, conferees felt that *all* scientific developments can be used to benefit or harm mankind. New techniques provide no more dangerous possibilities for the abuse than existing ones including artificial insemination, hormones, drugs, and behavioral techniques. Society has to protect itself against the misuse of *any* new findings, not by preventing their discovery, but by enforcing reasonable laws. When abuses become likely, scientists should inform the public and help legal experts and concerned laymen plan the proper laws. The NIH report concluded on this serious note: "At present the implantation of normal or allophenic human blastocysts seems to be close enough to reality so that new laws may be necessary."

Genetic Engineering

The prospect of "allophenic human blastocysts" leads us a step further. Dr. Beatrice Mintz, one of the conferees at the NIH meeting on genetic counseling in 1971, has for some years been experimenting with "mosaic" or "muddled" mice. These are produced by combining multiple fetuses into one. The first mosaic mice were born on New Year's Day in 1965. The outward appearance of these combinations is striking—vertical and horizontal stripes, round and irregular spots, different colors, and so on. There are of course other than outward differences in such a fusing of different creatures into one.

Robert Francouer, in his book *Utopian Motherhood,* humorously speculates on human beings' adopting this approach, with two mothers sharing an embryo, perhaps flipping a coin to see who will keep the finished product after each has carried it for four and a half months. Such a practice would curb the population explosion and reduce the orphanage problem, besides saving money for parents, according to Francouer. There are more serious implications in the experiments, of course.

Dr. Mintz has produced "allophenic" mice by assembling genetically different embryos and reimplanting the composites into surrogate mothers. These are "four-parent" mice, each with cells of two different egg origins. Cells with a genetic defect can thus be combined with genetically normal cells; such coexistence permits lethal genes to be "rescued."

Defects which have been partially or wholly corrected by this technique are retinal degeneration and severe and ordinarily lethal anemia. Mosaic retinas with partial function have been obtained from mixtures of defective and normal cells. In the anemia correction, it has been found that new blood defects may show up in the genetically normal red cells, an indication of the possible unexpected hazards in this form of rescue. Other allophenic strains of mice have been susceptible to malignancy, including liver, lung, and mammary tumors, and also leukemia. Some mice have been sterile or sexually abnormal. About half the embryos die.

If mouse cells will fuse, how about those of two different animals? It turns out that they will. Cell hybridization was first observed by French researchers Georges Barski and Boris Ephrussi in 1960. And in 1964 researchers at Carnegie Institution demonstrated that DNA in human genetic material has many sections that are identical with monkey DNA. As predicted by evolutionary theory, mouse DNA combined nicely with that of other rodents such as rats and hamsters, but showed much less attraction for the DNA of monkeys and cattle. Human DNA demonstrated only moderate attraction to that of the mouse, but it combined with some from a Rhesus monkey almost as strongly as if it had come from a human being. Both mouse and human DNA showed only weak attraction to DNA from salmon, and almost none to that from bacteria.

Man and mouse cells were first hybridized in 1967 at New York University. The main purpose of such work is to provide a medium for growing and observing single human chromosomes. But what about the genetic production of hybrids between animals, and ultimately between animals and *men?*

Artificial insemination has been used to produce hybrid crosses. For example, from the zebra and the mare, "zebroids" have been born. Cows and zebras have been crossed, as have cows and bison. Sheep and goat pairings have not succeeded, however; nor have dogs and foxes, or rabbits and hares. Bruce Wallace, a professor of genetics at Cornell University, says that the technology for breeding cattle and other animals has "simultaneously made gorilla-man and chimpanzee-man hybrids possible." He also suspects that man-ape crosses might succeed.

John R. Platt, biophysicist at the University of Michigan, is quoted in the journal *Science,* December 2, 1966, as follows:

It would also be useful to try animal copying with the nucleus taken from one species and the egg in which it was implanted taken from another. Donkey and horse can be mated. Will a donkey nucleus in a horse cell give a donkey or something more like a mule? This might teach us something about the developmental embryonic differences between species. If it would work we might be able to save some vanishing species by transplanting their cell nuclei into the egg cells of foster species. Is the DNA that carries heredity destroyed immediately when an animal dies? If the meat of wooly mammals locked thousands of years in the arctic ice is still edible, perhaps their DNA is still viable and might be injected, say into elephant egg cells to give baby wooly mammals to the elephant. By some such methods, perhaps we might achieve "paleo-reconstruction" of the ancient Mexican corn or mummy wheat, or even the flies that are sometimes found preserved in amber.

Discussing the possibility of the existing experimental work with hybrids, the mixing of man and mouse chromosomes, Joshua Lederberg says: "Before long we are bound to hear of tests of the affect of dosage of human 21st chromosome on the development of the brain of the mouse or the gorilla."

For some, the greatest potential in genetic engineering is not in negative techniques that eliminate harmful traits, but in positive engineering to create better people. Rather than therapy, why not *improvement,* or complete change in some cases? Jean Rostand suggested the chemical triggering of increased cell division in the brain, thus increasing the size of a brain before birth and producing a supergenius. Such a superbrain was foreseen by Renan in 1871. Of course it would necessitate Caesarian section, or even the artificial womb.

Next to the Liners stood the Matriculators. The procession advanced; one by one the eggs were transferred from their test-tubes to the larger containers; deftly the peritoneal lining was slit, the morula dropped into place, the saline solution poured in . . . and already the bottle had passed, and it was the turn of the labellers. Heredity, date of fertilization, membership of Bokanovsky Group—details were transferred from test-tube to bottle.

Aldous Huxley, *Brave New World*

10 Tempest in a Test Tube

In 1932 Aldous Huxley's *Brave New World* was an immediate success. So enduring was this success that it is even more timely forty years later. Huxley missed on several points: there is no mention of atomic bombs or nuclear energy, "sleep teaching" is described as a potent force in shaping his society, and the time scale of his predictions errs mightily. In *Brave New World,* the Central London Hatchery and Conditioning Centre is decanting its mixed bag of planned citizens in the year A.F. (After Ford) 632. By 1946, when he revised the introduction to the book, Huxley had reduced the time to his vision of the future to only 100 years, but even that estimate was far too conservative. The genetic engineering he feared is no longer fiction.

A speaker at a recent genetics conference suggested that his colleagues reread *Brave New World*, which they would find far less revolutionary than it had seemed in the thirties. As a matter of fact, parts of the book sound more like a report of what is happening on the genetics front than a cautionary tale about a controlled society. Indeed, about all that needs doing to make it a factual report is to change the fictional names Huxley used to those of the biologists who are actually doing these things in genetics laboratories.

The first prescriptions for "test-tube babies" were written centuries ago when straight-faced "scientists" set down recipes for creating "homunculi" from sperm cultivated in a jar. The idea has continued to nag at us since then, first on the fictional fringes, then in scientific papers, and finally in the laboratory where the factual forerunners of Huxley's decanted Alphas and Betas are being hatched.

The Ethereal Copulators

By the time Huxley wrote of his "bottlers" and "matriculators" on the hatchery assembly line, artificial insemination already had a long history.

In 1884, ten years before Huxley was born, Dr. William Pancoast engineered what was probably the first artificial insemination. Without the knowledge of either the husband or the wife, Pancoast and his colleagues artificially inseminated an anesthetized woman with semen from one of the student doctors at the hospital. A healthy child was born. Pancoast later confided in the husband, who apparently was not disturbed and only asked that his wife not be told. As word of the new technique leaked out, Dr. J. Marion Simms admitted that he had arranged "ethereal copulation" 55 times, and claimed a success rate of about 4 percent. (Today artificial insemination is successful about 20 percent of the time.)

Genetic Bastards?

According to an article in the *Medicolegal Digest* in 1961, there were an estimated 50,000 "proxy babies" in the United States, with more coming on at the rate of 1,000 to 1,200 a year. Despite this popularity, the question of the legality of artificial insemination was described as "a legal vacuum." Among the questions that could not be answered: Are such children legitimate in the eyes of the law? May the mother's husband or the donor's wife sue for adultery? May a husband later disclaim the child and refuse to provide for it? May the AID (artificial insemination donor) child, discovering his true paternity, claim heirship to the donor's property? Can the child's relatives deprive him of inheritance? In London, the *Medicolegal Digest* article said a couple had their child branded illegitimate by court ruling. A New York physician was accused of rape for providing artificial insemination.

Today an estimated 10,000 AID babies are born yearly in America, although some writers have suggested that there may be this many births in New York City alone. Yet little clarification has come in the last decade. Artificial insemination is still largely ignored by the legal apparatus, and most states have not given status to this kind of conception. Some states consider adultery a criminal act, and artificial insemination may possibly fall under this heading.

Even more remarkable than the legal vacuum surrounding artificial insemination is the reaction of the public to this sidestepping of the normal procreative act. A Harris poll taken in 1969 showed that only about 3 percent of the American public had even heard of artificial insemination! It is interesting to speculate on how many of the 150,000 or so AID Americans know either the technique or the fact that they were conceived in this fashion.

The Sperm Bankers

Hermann Muller, who spent much of his life pleading for the establishment of sperm banks, would be pleased to know that such banks are now available, where men can leave sperm samples against the day they wish to withdraw them for implantation. This practice is said to be popular with men about to undergo vasectomy, or with those who plan to have children later in life but want germinal cells produced in their most vigorous and healthy years. There are still the mercenary donors of sperm for a fee.

We have not yet reached the genetic sophistication of the James Costigan play *Baby Wants a Kiss,* in which women have children via "celebrity seed" donated by sports stars, movie actors, or other "greats." However, there seems little to stand in the way of Muller's dream of the sperm bank set up by contributions from the leaders of our time—except to be sure which "model" types we want to reproduce. The technical details seem easier to solve, and many have already been worked out.

The freezing of semen was first attempted by Lazzaro Spallanzani, probably about 1776. However, not until 1949 did Dr. A. S. Parkes of London's National Institute use glycerol as a protective agent which permitted sperm to be successfully stored for months at a time. With this capability, "telegenesis" or reproduction at a distance became an inexpensive and useful technique. Frozen human semen was first used in 1958 by two Japanese doctors, and samples frozen for more than 18 months have been used successfully in artificial insemination. It is believed that semen could be effectively stored over indefinite periods of time. Specimens taken from bulls have been kept for more than ten years, apparently with no genetic damage or loss of vitality. Germinal cold storage has been suggested for guarding against radiation damage in the coming decades. Nuclear physicist Ralph E. Lapp is among those who have advocated the storage of semen in lead-shielded vaults to assure an undamaged supply of germinal tissue. One biologist has noted that with sperm banks an established fact, banks of frozen ova cannot be far behind. One step further along the Hatchery assembly line is the "embryo bank." None of these techniques is any longer fictional.

Artificial Inovulation

While philosophers ponder and equivocate, genetic engineers are moving ahead rapidly and on several fronts. The telegenesis idea has been expanded from the storage and transport of semen to similar techniques with eggs. "Superovulation" experiments have produced more than 100 rabbit eggs in one female, and fertilized "superrabbits" are shipped halfway around the world, where the eggs are removed and implanted in other rabbits. Such techniques in animal husbandry add to the prize bull a prize cow to produce fertilized supereggs which could then be transplanted into normal animals for gestation.

Huxley's *Brave New World* geneticists boasted of producing thousands of eggs from a single womb. While such mass fecundity has not yet been demonstrated, modest superovulation has. The human ovary is about the size of a large almond, but it contains potentially about half a million eggs, of which only about 500 will be released during a woman's fertile period of life. Contraceptive pills, given and then withheld, have resulted in superovulation in women. In 1968 a woman in England was prescribed a gonado-trophic hormone. Childless at age thirty, she surprisingly gave birth to *sextuplets*—four girls and two boys. Hormone-induced quintuplets were born at about the same time in Sweden, New Zealand, and New York City.

Dr. E. S. E. Hafez, chairman of the Department of Animal Sciences at Washington State University, is enthusiastic about these advances in artificial insemination, telegenesis, and the like. In 1966 he suggested that within ten or fifteen years a woman could shop for her offspring from an assortment of packets much like those used for flower seeds, containing already fertilized eggs— frozen, one-day-old embryos with a pedigree including sex, hair color, size, IQ, and even a picture of how the child would probably look. If desired, the shopper could have this embryo implanted in her own uterus, but there would be little reason for this old-fashioned procedure, as we shall see.

Dr. H. Bentley Glass also foresees "prenatal adoption," with selected embryos implanted in the foster mother's uterus. By checking sperm and egg donors with a battery of biochemical tests, matings of carriers of the same defective genes could be avoided, or defective embryos could be detected and discarded. Couples would be able to store their own fertilized egg (conceived early in marriage) for later reimplantation when they wanted to have the child. This would minimize the risk of congenital defects that tend to occur more frequently in children of older couples.

Late in 1972 scientists at the Atomic Energy Commission's Oak Ridge National Laboratory in Tennessee announced that "deep-frozen" mouse embryos had been used to produce living, and apparently normal, mice. In the experiment more than 2,500 mouse embryos were recovered from naturally pregnant mice and slowly frozen with liquid nitrogen to temperatures as low as 452° below zero Fahrenheit. After as long as eight days at this temperature, the embryos were slowly thawed. At normal temperatures once more, between 50 and 70 percent of them developed in test tubes to the "blastocyst stage," the point in cellular development considered to be the beginning of fetal growth. One thousand such fetuses were then transferred into the wombs of mice, and about 65 percent of these "host" mothers accepted the pregnancies. More than 40 percent of the transplanted embryos resulted in full-term mice, apparently normal in all respects.

Oak Ridge scientists pointed out that the deep-freeze technique might provide a safe and reliable method for long-term preservation of human hearts and other transplantable organs. It was also suggested that success of the technique with larger domestic animals would "facilitate worldwide dissemination of stock with an optimal genetic background for a particular use of geography." There was also a careful disclaimer by one scientist that he doubted seriously that the mouse experiments could ever be applied to the production of human "deep-freeze babies." There are, of course, those who are not nearly so doubtful.

At Cambridge, England, Drs. Robert G. Edwards and Patrick Steptoe are leading the work in artificial inovulation. Dr. Edwards began his work in this field at Johns Hopkins University in 1965 with pig ova, and by 1969 was artificially inseminating human eggs. It is believed by informed scientists that the first "ectoconception" baby will arrive shortly. Some fifty women in their mid-thirties, all sterile because of oviduct blockages, are said to have volunteered for the operation, in which eggs are surgically removed from the woman and then artificially fertilized in a culture dish. After more than a hundred cell divisions, the developing embryo will be reimplanted on the uterine wall and normal gestation will take place, if all goes well.

In justifying such experiments, Dr. Steptoe says: "If you could see, as I do every day, the anguish of couples who want babies, you'd want to do everything possible to help them."

The Surrogate Mother

The facility with which scientists now seem able to remove and replace embryos leads to the humorous suggestion that mothers-to-be will be able to take the baby out to see how it is progressing. It also suggests that the artificially fertilized egg need not be put back inside the original mother at all, but in a "surrogate" mother, hired for the occasion. One scientist has mentioned "wombs for rent," an extension of the age-old wet-nurse idea. There is no immune response in the womb, so such transplantation would be feasible.

Robert Francouer raises the peripheral problem of the effect of smoking which the host or "hostess" mother would have on the child. Children born of smoking mothers generally weigh six ounces less than those of nonsmokers. Francouer refers to the ethics of the wet nurses established "long before the advent of Pablum and Similac." These women were held to be the epitome of Christian charity, but would the surrogate mother fall into the same category? While some might applaud the sparing of a mother for nine months, there would still be the female surrogate paying the price. Even militants advocating virgin birth might prefer gestation outside the female womb.

The Artificial Womb

It has been noted that women are the only mammals who have their offspring in intense suffering, running the appreciable risk of death. As a result, some suggest that if women cannot have a painless and riskless birth then perhaps the solution is the artificial womb.

The artificial womb is a reality; one has been developed by the National Heart Institute. A chamber filled with synthetic amniotic fluid and connected with an oxygenator for fetal blood has kept lamb fetuses alive for two days. Such "ectogenesis," or development outside the natural womb, is hailed as an important and logical extension of artificial insemination. An enlightened female should not put up with nine months of physical suffering to produce a child, one argument goes. Procreation, considered by some as beneath human dignity in the "animallike" way it is traditionally done, would be replaced by clinical techniques for joining sperm and egg in an artificial womb, and would take place where it could be scientifically nourished and carefully monitored to guarantee a "perfect product."

We have long had incubators that take over when needed with

premature infants born at five months and earlier. The artificial womb is technically called an "extracorporeal membrane oxygenator." It has not yet been used to save the life of an early human embryo, although a human fetus survived for six hours. Other laboratories are working on different versions of the artificial womb.

Blood transfusions are successful, along with kidney transplants, anesthesia during childbirth, and the artificial feeding of babies. But embryologist Francouer seriously questions the approach of substituting for nature for the whole nine months of gestation. Could an artificial womb be made that would be completely harmless to the embryo? What would be the psychological effects of keeping a human fetus for nine months in "a womb of glass and steel"? Francouer points out that five of thirteen heart transplant patients at Stanford University became psychotic, while three others showed signs of organic brain damage. Would there be similar side effects with artificial wombs and surrogate mothers?

Ectogenesis

". . . take ectogenesis. Pfitzner and Kawaguchi have got the whole technique worked out. But would the governments look at it? No! There was something called Christianity. Women were forced to go on being viviparous" (*Brave New World*).

The term "test-tube baby," in use for many decades, describes the product of artificial insemination. The semen has perhaps been stored for some period of time, but the mechanics of conception takes place somewhat as it would normally. "Ectoconception" moves away from this natural process. "Ectogenesis," a feature of Huxley's "Hatchery," is the next logical step, and now we are properly entitled to use the term "test-tube baby," for in this process egg and sperm are united and allowed, or coaxed, to grow in glass or steel rather than a natural womb.

Many researchers, in addition to Edwards and Steptoe in England, are performing ectogenesis experiments. Landrum Shettles has done such work in America, and Daniele Petrucci, an Italian geneticist-biologist, claims to have kept a test-tube fetus alive for 59 days. It died then owing to a technical mistake. In 1966 Russian scientists announced that they had managed to keep more than 250 human embryos alive beyond Petrucci's record of 59 days. One fetus was reported to have lived 6 months and to have reached 1 pound 2 ounces before dying.

The question was raised in the Vatican, against Petrucci, whether or not ectogenesis is permissible. In either case, the next question was whether termination of such life is abortion, murder, or some other criminal act. There is great controversy in this area. Father Joseph Donceel, a Belgian Jesuit teaching at Fordham University, has stated: "I feel certain that there is no human soul, hence no real human being, during the first three months of pregnancy, and I consider as safe the norm which applied in medieval Catholic England that the Common Law permitted abortion when it was performed before quickening."

Another view on the beginning of human life was reached during the First International Conference on Abortion, held in Washington, D.C., in October, 1967. This meeting was attended by some sixty authorities in medicine, law, ethics, and the social sciences. The conclusion of the medical group was:

The majority of our group could find no point in time between the union of sperm and egg, or at least the blastocyst stage, and the birth of the infant at which point we could say that this was not a human life. . . . the changes occurring between implantation, a six week embryo, a six month fetus, a one week old child, or a mature adult are merely stages of development and maturation.

Theologian Paul Ramsey is one of the most outspoken against the genetic experiments described in this chapter, and he considers them a perversion of the great gift of creation: "I assume that man will refuse to live in a crystal palace, or devote his every purpose to becoming a 'card-carrying cadaver.' Man was not born from a chemist's retort, the scoundrel. To the chemist's retort he will not return. God or no God. Crick or no Crick." The Crick referred to is Francis Crick, the physical geneticist, whose views are considerably different from Ramsey's.

> . . . a bokanovskified egg will bud, will proliferate, will divide. From eight to ninety-six buds, and every bud will grow into a perfectly formed embryo, and every embryo into a full-sized adult. Making ninety-six human beings grow where only one grew before. Progress. . . . Ninety-six identical twins working ninety-six identical machines!
>
> Aldous Huxley, *Brave New World*

11 "Clones": Xeroxed Human Beings

More than 200 years ago Lazzaro Spallanzani began to study the regeneration of limbs in some animals and asked, "Should the expectation of obtaining this advantage for ourselves be considered entirely chimerical?" The dream was even then an old one, based on the observation that many amphibians have the power of regenerating parts lost by accident or experiment in the laboratory. An earthworm cut in two grows into two complete worms; indeed some species normally reproduce in this way. Genetic engineering might well include the ability of man to regenerate lost body parts. However, there is another prospect that makes this seem a very modest wish.

A science-fiction story several decades ago told of a man going to a hospital for surgical removal of the tip of his finger. Years later he was astonished to meet *himself,* in an identical copy "grown from the finger"! The basic idea in the science-fiction yarn and the actual process of "cloning" is this: If a body can grow a new finger, then a finger—or just one cell—can grow a new body. Dr. J. B. Gurdon accomplished the feat in Oxford University in 1966. He didn't clone a man, but he did produce a host of identical African clawed frogs by a process more like setting out cuttings from plants than animal reproduction.

Joshua Lederberg, who shared the 1958 Nobel Prize in medicine and physiology for his work in genetics, recently was quoted on the implications of Gurdon's work with amphibians: "There is nothing to suggest any particular difficulty about accomplishing cloning in mammals or man, though it will rightly be admired as a technical tour de force when it is first accomplished."

Parthenogenesis: Virgin Birth

Ever since the sperm was first identified, the male sex cell has generally been considered the *sine qua non* of reproduction in sexually reproducing species. However, a number of experimenters have for some time sought to jostle nature toward reproduction without such a conception trigger.

Parthenogenesis is reproduction through the development and growth of an unfertilized cell. (The Parthenon was the temple of the virgin goddess Athena.) Parthenogenesis occurs naturally in plants, and also in some insects, such as the drone bees that result from unfertilized eggs. In 1896 German embryologist Oskar Hertwig and his wife, by adding strychnine or chloroform to sea water containing sea urchin eggs, fertilized them without any contact with sperm. Three years later, another German, Jacques Loeb, repeated these experiments with the successful chemical fertilization of sea urchins. Loeb later came to work at Woods Hole Marine Biological Station at Cape Cod.

Further experiments yielded parthenogenetic frogs, and in 1939 Gregory Pincus produced such a rabbit. This outwardly normal rabbit, born as the result of thermal shock treatment, was featured on the cover of *Life* magazine. Pincus and his associates, however, were working on contraceptive techniques and were not interested in pursuing the parthenogenetic experiment per se.

In 1952 Dr. A. D. Peacock summarized research on parthenogenesis at a meeting in Belfast, and reported 371 different ways of producing sea urchins without the help of a father! These included 45 different kinds of physical shocks ranging from simple shaking in a glass vial to thermal shocks, 93 chemical methods, 64 biological methods, and 169 combinations thereof.

Dr. Helen Sperway, lecturer on genetics at London's University College, in 1955 suggested that one in every 1.6 million pregnancies would involve a spontaneous and natural doubling of the egg's chromosomes, resulting in a parthenogenetic birth. And under the headline "Life Without Father," the *Manchester Guardian* reported the following human case of virgin birth under study by British scientists.

In 1944 a young German woman collapsed of extreme exhaustion during a bombing of Hanover. Although she claimed not to have had sexual relations, she became pregnant and gave birth to a daughter. (Parthenogenetic mammals must always be females, since there is no Y chromosome involved in their heredity.) The English investigators found no evidence of hereditary differences between mother and daughter. Genetically, the two appeared to be identical twins. However, a skin transplant, ordinarily successful in such cases, failed to take.

There is a genetic difference between the female from whom the egg comes and the product of such an unfertilized egg, however: the egg cell is haploid, containing only half a set of chromosomes. Human sex cells contain only 23 chromosomes rather than the full complement of 46, and a parthenogenetic offspring is thus not an exact genetic copy of its parent. A normal animal is diploid, possessing a double set of chromosomes with half of them from each parent. Two Swedish researchers have artificially produced "triploid" rabbits. In this mutation the reproductive cells are treated with a chemical and the fetus acquires three sets of chromosomes rather than just two. Using this technique, Japanese geneticists have produced a triploid watermelon which is seedless.

Cloning: "Chips off the Old Block"

The book *Evolution and Theology,* published in 1931 by Dr. Ernest Messenger, expressed the following idea, of particular interest to those pursuing novel methods of reproduction: "In principle, every cell in an ordinary organism contains at least radically the virtuality of the species and the race." Messenger intentionally said "every cell," instead of just the sex cells. It was not long before someone else took up this unusual line of reasoning.

In 1938 Nobel Prize German zoologist Hans Spemann suggested an experiment which even to him seemed "somewhat fantastical." His proposal was to remove the nucleus from an egg and replace it with the nucleus of some other cell, preferably an adult cell. "This experiment might possibly show that even nuclei of differentiated cells can initiate normal development in the egg protoplasm."

Fantastic or not, by 1952 Drs. Robert Briggs and Thomas J. King at the Institute for Cancer Research in Philadelphia replaced the nuclei of freshly fertilized egg cells of the leopard frog, *Rana pipiens,* with the nuclei from early embryonic tissue of a single individual of that species. They produced a school of free-swimming tadpole embryos, all having the same genetic endowments as the tissue cell donor. In 1956 embryonic tadpoles were produced, from older donor cell tissue, and a few of the individual duplicate tadpoles were allowed to grow to maturity. Remarkable as this work was, it did not yet use an adult or fully differentiated cell to "initiate normal development in the egg."

True cloning began with the work of Dr. F. C. Steward, director of the Laboratory for Cell Physiology at Cornell. A decade ago Steward tried stimulating the growth of *differentiated* cells in carrot roots. Normally, it is egg cells that grow. The process of fooling eggs or seeds into growing without being fertilized is nothing new, and a variety of stimuli will serve: physical contact, heat, light, and so on. But when Steward caused differentiated cells to grow into more carrot plants, something new had been added to the techniques of the plant geneticist. Chrysanthemums are now produced commercially by cloning. Far bigger things were coming, although it has been said that growing a carrot from a differentiated cell is itself basically more startling than the extension of the process to cloning a man.

A carrot is not anything like a man, however, and there were difficult intermediate steps to be taken. With an eye on the Briggs and King experiments, a British scientist began to move up these steps. By 1961 Dr. Gurdon, mentioned earlier, produced a batch of offspring from the South African clawed frog, *Xenopus laevis,* each having exactly the same genetic characteristics of the cell donor. However, critics pointed out that this wasn't cloning but simply the renucleation of an already fertilized egg cell.

In 1966, starting with an unfertilized egg cell, Dr. Gurdon prepared it for true cloning by radiating its nucleus with ultraviolet light. He then replaced the destroyed nucleus with one taken from an intestinal cell of a tadpole of the same species. Endowed with a double set of chromosomes, rather than the single set of its unfertilized stage, the egg cell began to grow as if it had been fertilized normally. Cloning had been accomplished in an animal.

Here is the key point of cloning: the egg cell is essentially an environment; it is the implanted nucleus with its full set of chromosomes that determines the genetic makeup of a cloned individual.

The Asexual Revolution

Dr. J. B. S. Haldane was long ago convinced of the possibility of cloning humans and welcomed the notion even while most of his scientific brethren scoffed at the idea of immortalizing anything more than a patch of skin through tissue culture in the laboratory. For example, Haldane proposed that the donors in clonal reproduction should spend their time after age fifty-five teaching their carbon copy offspring. The guffaws increased, but when Gurdon made it all seem about to come true, there was a sudden crush to get aboard the biological bandwagon.

England's Lord Louis Rothschild says, "The cloning of humans is a near possibility soon to be realized." Caltech's James F. Bonner in 1971 predicted that human mass production would be possible within fifteen years. By then it will be possible to order carbon copies of individuals, and thousands of latter-day Einsteins and Beethovens could be produced, along with thousands of "typical assembly-line workmen." Since carbon is such a vital ingredient of life, the "carbon copy" analogy is especially apt, but "Xeroxing people," a catchier phrase, quickly gained currency.

Dr. Elof Axel Carlson of the University of California at Los Angeles went even further:

I predict the synthesis of the human genotype. The techniques for introducing this into nucleated fertilized eggs will also be developed. This will permit the necrogeneticist to bring back individuals (e.g. historical personalities) of identical genotype to the dead by using the sequences worked out from the entombed tissues. The resemblance of these individuals to ancient photographs and paintings will be startling, but their personalities will be no more like their predecessors than are identical twins to one another.

Fanciful and far-fetched as the resurrection of King Tut appears, other proposals make even this man-made second coming seem mild. Some years back the notion was advanced that dead people be frozen and held in reserve until science developed a cure for the cause of their death, including old age, presumably. Here is a specter for the Zero Population Growth advocate to ponder: billions of bodies, stacked like cryogenic cordwood awaiting resuscitation, while in nearby freezers similar billions of selected day-old embryos wait to be born! Perhaps the only population solution then will be a long freeze settling over the whole world as a moratorium on the ceaseless production and preservation of life on our planet.

Dr. Robert Sinsheimer, director of the biology division at Caltech, and one of those at the forefront of the genetic revolution, has said that a human clone will be produced by 1980. Other scientists have advanced a variety of reasons why this cannot be. For example, the frog eggs Dr. Gurdon used are very large eggs, while mammalian eggs are tiny and thus more difficult to manipulate. However, Dr. John Watson points out that there are existing techniques (developed in entirely different connections) in which mammalian cells can be fused, thus making possible a cloned mouse. How big a step is it then from mouse to man?

Watson points to the fact that much research necessary for cloning people is going on in other areas at present, speeded by concern over population, hereditary diseases, and other problems. He too foresees the production of a human clone within a generation, and deplores the lack of concern by government and individuals thus far concerning this prospect for the biological revolution. Perhaps there has been too much crying of wolf, too much fictionizing, for the man in the street to get worked up or even waked up about what is actually going on in the laboratory.

The Xeroxing of Excellence

The implications of cloning are obvious. Could we use another Einstein or George Washington? Would the music of a dozen

Tchaikovskys, the paintings of teams of Leonardos benefit the world? How valuable would a dozen Man O' Wars be in horse breeding? The poignant possibility of replacing human beings lost to illness or accident suggests another use for cloning (although some consider grisly the thought of "clonal farms" where duplicate copies could be kept "in pasture" or on ice against the need of repair parts).

It has been suggested that cloning, at its best, might permit the copying of the finest genetic combinations that arise in the human species—just as the coming of writing made possible the copying of the fruits of great human minds. On the other hand, what might another Hitler do with hordes of identical "super-Aryan" clones!

Somewhat surprisingly, theologian Paul Ramsey lists a number of advantages favoring cloning. The first is the speed of the process: there would be no need to wait a generation to see whether or not a genetic engineering experiment had worked. It would work, because the clone is an exact copy of the individual it is produced from.

The second advantage is that parents could copy themselves. Ramsey points out that cloning could allow parents to produce an exact copy of either one of them without the danger of mixing harmful recessive genes. Nevertheless, he disagrees with Haldane's view and believes that mixing the parental relationship with the twin relationship could produce a narcissism psychologically disastrous for the young Xerox copy.

Third, according to Ramsey, clonal reproduction might have considerable social advantages. It would be a procedure for making humans with interchangeable parts, because the members of a clonal colony would all be identical twins, and thus able to freely exchange organ transplants with no danger of rejection. "Intimate communication among men" would be notably increased. Misunderstanding would be lessened. There would be better social cohesion and cooperation. This expectation stems from the belief that identical twins are especially sympathetic and easily able to interpret each other's minimal gestures and brief words. We might produce pairs or teams of people for specialized work in which togetherness is a premium—astronauts, deep sea divers, surgical teams, and so on. Clonal reproduction might even cure the generation gap!

A fourth advantage is the scientific appeal of cloning. We would be able to see if a second Einstein could even outdo the first one; although, as Ramsey points out, this would take more than twenty years to determine.

To Clone or Not to Clone

We may already be undergoing a sexual revolution in which love and procreation are physically separated. Cloning and ectogenesis would seem to hasten this process, and completely separate romantic involvement from the production of new life. What effect this will have on parents (or parent), and more importantly on the resulting children, is difficult to foresee. Normal heredity and the genetic process of natural selection would seem to be done for. However, many feel that natural evolution has in effect already ceased to be an appreciable factor in man's development, that we have been engineering our own evolution for a long time through selective marriage, social customs, nationalism, and technology— including the radiation from nuclear bombs and power plants. If nurture is the factor that environmentalists believe it to be, society overrides the genes in another way anyhow.

Be that as it may, there are many who are afraid that the potential in clonal reproduction is anything but good. Dr. Willard Gaylin, president of the earlier mentioned Institute of Society, Ethics, and the Life Sciences, voices such fear. In an article titled "We Have the Awful Knowledge to Make Exact Copies of Human Beings," he deplored the reaction this knowledge produced in nonscientists: "more titillation than terror, visualized as a garden of Raquel Welches, blooming by the hundred, genetically identical from nipples to fingernails."

Gaylin argues against the probability of producing multiple copies of great scientists, athletes, or warriors, because environment plays such a vital part in shaping the hereditary material. St. Francis, he says, could have become a tyrant and Hitler a saint.

The fact is that they did *not,* however; and those reflecting on long and futile attempts to rehabilitate the antisocial might see more hope in a thousand genetically identical St. Francises than a thousand genetically identical Hitlers. Gaylin also cites the argument that identical twins (which he describes as "nature's clones") may develop very differently both physically and socially. Statistics show that identical twins are far more alike in such things as appearance, height, weight, and even psychical temperament than are other people. However, Dr. J. B. Rhine, the ESP researcher, reported this information concerning twins: "Nothing outstanding has occurred during any single case of identical twins tested so far. The averages on the extra sensory test were approximately the same whether the sender and receiver were identical twins, paternal twins, singleton twins, singleton siblings or simply friends."

There are other objections to the clone. Joshua Lederberg suggests that clonality would be an "evolutionary cul de sac" that would multiply a rigidly well-adapted genotype to fill a stationary niche. As long as the environment remained static, cloned humans would congratulate themselves on their short-term advantage. Should it change, however, they would be in trouble. The clone *would* be a genetic dead end as far as evolution of the first order is concerned. Would clones be able to mate among themselves; would they even want to revert to this animal practice? What sort of progeny would come from pairings of clones and normal humans?

Lederberg also comments on the narcissistic motives that would impel a clonist, and of course this trait would be passed on to the clone. Clonal reproduction, then, could make for clannishness. Lederberg advocates "tempered clonality," a combination of asexual and sexual reproduction. Sexual reproduction would be used for experimental purposes, and when a suitable type was produced it would be maintained by clonal propagation.

The geneticist would not have to stand by and watch clone beget identical clone ad infinitum, of course. Scientists have manufactured DNA and altered the hereditary makeup of bacteria. In theory, this could be done with human organisms; it is a matter of degree. With tailored nucleic acids human beings might be produced with more brains (or less), and more or less muscle, fat, limbs, hair, and so on. Arthur C. Clarke has suggested that legless astronauts would take up less space and require less propulsive power to get them to distant stars and back. They might also be engineered to be ageless, so they could complete such long journeys in one "generation." Having occasionally retooled a desirable type, geneticists could then clone away as needed.

Cloning suggests the need for but one sex for reproduction. Women could rule their own world. So could men. "Unisex" clones might be produced, or even neuter drones like those in bee society.

Robert Francouer likens the creation of Eve from a portion of Adam to cloning. However, he suggests that the first cloned man, the new Adam (or Eve), would be an orphan in an especially poignant sense. The clone would be truly a child of the race, selected and produced by its collective wisdom. But how would it fit into our ongoing society, how would it be received both by us and by its genetic doubles?

Life is beginning to cease to be a mystery, and becoming practically a cryptogram, a puzzle, a code that can be broken, a working model that can sooner or later be made.

J. D. Bernal

12 The Artificial Creation of Life

Much is written and said about the "creation of life," and surely parents like to feel that they have indeed produced new life. However, while we as mothers and fathers pass on something of ourselves to our children, we do not *create* life in the process. We simply transmit it, much as Olympic runners pass on the torch. The spark of life resides in the nucleic chemicals of the nucleus, the genes and chromosomes that blueprint and drive the cells in their work. When the Creator produced the first living thing on earth, He set in motion the ongoing process called life.

When we strike a match, we are not inventing fire but merely initiating a physical and chemical process that is inevitable under certain physical conditions. Life is a similar flame with an important exception: it has never gone out. Indeed, some who have traced the origin of life feel that should the spark ever die out or be extinguished it might never reoccur: that the conditions that led to its production in the original instance no longer prevail. Lucretius put it: "Mortals live by mutual interchange. One breed increases at another's decrease. The generations of living things pass in succession, and like runners in a race they hand down the torch of life." But that torch is far more precious than those the ancients guarded with their lives lest they flicker out with the dying flame.

There are primitive people today who do not possess the skill of making fire with flint or friction. Instead they must carefully guard existing fires, keeping precious embers burning all day, and all year, indeed down through the generations, so that each may have the warmth and other benefits of the process called combustion. Life is somewhat like the process of burning; metabolism is much the same sort of physical process of combining a fuel with oxygen to produce energy to sustain life. Mankind and the animals seem to possess inherent knowledge of the preciousness of the spark they harbor, and the law of preservation of the species overrides all others. The durability of the genetic stuff aids in safeguarding life. In fact, it is easy to entertain the notion that immortality resides not in replicated men or animals, but in the genes themselves. As some have said, life goes on to keep the gene alive. If there is an élan vital as Bergson and others have argued, this force must reside in the giant DNA molecules of the gene.

What Life Is

It is generally agreed that life has three fundamental properties: metabolism, growth, and the power to reproduce. Metabolism is the chemical process in a living cell in which "fuel" is converted into the energy needed for life to proceed. Growth is self-explanatory; so is the power of reproduction.

Some people (generally poets, but others as well) make cases for clouds, oceans, and such as living things. A cloud, by some stretching of the imagination, can be thought of as exhibiting metabolism, growth, and even reproduction, since the moisture from a dying cloud may "seed" a new one. The ocean as a living thing is another apt example, since it exhibits movement, chemical and physical processes, and the ability to produce other bodies of water. We speak of water "poisoned" by pollution and dying (or eutrophying) with old age. An even more convincing "living thing" is a crystal. Some crystals grow, repair themselves, and seed new growth. Geneticist Jacques Monod suggests the crystal as the closest inorganic approach to what we generally consider living things.

Man has produced machines that exhibit properties comparable to living processes. Factories with automated machines and computer controls are analogous in some respects to living things. An automobile is a substitute horse or other beast of burden; an airplane rides the same air that birds do. Automobiles don't have little automobiles, nor do airplanes produce their own progeny, but there has been speculation about building machines with the power to repair themselves and to produce more machines. It might even be possible to build into them the ability to evolve in response to changing environment. Whether success would qualify such robots as "living" is a good question.

Another requirement generally included in the lists that separate the living from the nonliving is the distinction between "organic" and "inorganic." Swedish chemist Jöns Jakob Berzelius divided chemistry into organic and inorganic branches shortly after the beginning of the nineteenth century. An organic substance by his definition was something produced of living matter. Carbon seemed to be the vital element of life; all living things contain it. So do some nonliving things, diamonds, for example. But diamonds are associated with coal, and coal was at one time a living plant.

Carbon dioxide contains carbon, but this gas is the waste product from metabolism (and the combustion of engines, interestingly).

Generally accepted is the thesis that some point in past time marks the creation of life from nonlife, either by God or by blind chance. From that time on there was a definite and unchangeable distinction. A rock is dead, and so is a drop of water (although it may contain tiny living things). A leaf is alive, and so is a man—alive in a far more subtle sense than a phonograph or a factory. Except for man's insatiable curiosity, this division might have been satisfactory. Within two decades after Berzelius pronounced that living is living and dead dead, a student of his, Friedrich Wöhler, synthesized urea, an organic compound, from ammonium cyanate. The wall around life had been breached. The battle was joined, and it still goes on.

The Age-old Dream

For centuries man has dreamed of creating life. We have noted the measure suggested by the sixteenth-century pseudoscientist Paracelsus, who wrote a prescription for nourishing sperm into a "homunculus" or tiny man. Such male parthenogenesis is an interesting speculation, since the result would be "haploid" in its genetic structure. But it would not represent the creation of life from scratch.

In 1818 an English novelist wrote a book based on the artificial creation of life. Just how appealing—or horrifying—this prospect is should be indicated in the durability of the book's title: *Frankenstein*. Time has blurred the distinction, so that nowadays many associate the name with the monster rather than the doctor who created him. Frankenstein is nevertheless a household word and one that has probably spawned more "B" movies, lurid comic books, and TV shows than any other.

Mary Wollstonecraft Godwin Shelley, writer-wife of the great poet, was hardly creating a new idea in Frankenstein. Although many writers (notably including Jules Verne) are credited with prophecy, they generally put into fiction ideas already advanced by scientists of their time or even before. For example, Robert Fulton had invented a submarine, sailed it under the sea, and tried to sell it to France before Verne wrote his classic *Twenty Thousand Leagues Under the Sea*. No mad scientists had succeeded in creating a homunculus or living robot in Mary Shelley's time, of course, but there had been nervous fears of such creatures for centuries. Indeed, Jewish legend has many stories of golems, awesome artificial men.

It is generally accepted that Mary Shelley was writing an allegory rather than a straightforward tale about a real creature. She is said to describe the advent of science itself, and to foretell the stormy path that it and scientists themselves would tread. (Ironically, science and the citizen have collided in their most forceful encounter just at the fruition of the biological revolution.) However, it is interesting to go to the book itself for enlightenment.

In the preface the author states, "The event on which this fiction is founded has been supposed by Dr. Darwin and some of the physiological writers of Germany as not of impossible occurrence." This was Erasmus Darwin, of course, grandfather of Charles. Indeed, the twenty-one-year-old author often listened to her husband and Lord Byron discussing the origins of life and of reputed experiments of Darwin, "who preserved a piece of vermicelli in a glass case, till by some extraordinary means it began to move with voluntary motion." Mary Shelley thought perhaps "galvanism," the newly discovered force of electricity, might be involved.

In her story she described Dr. Frankenstein studying the ancient works of Cornelius Agrippa, Paracelsus, and Albertus Magnus. When he discovers the secret of life among the decaying corpses of charnel houses, Frankenstein says, "I was surprised that among so many men of genius who had directed their inquiries toward the same science, that I alone should be reserved to discover so astonishing a secret." In the preface, Mary Shelley wrote: ". . . I saw—with shut eyes but acute mental vision—I saw the pale student of unhallowed arts kneeling beside the thing that he had put together. I saw the hideous fantasm of a man stretched out and then, on the working of some powerful engine, show signs of life and stir with an uneasy half-vital motion."

A century after Mary Shelley, Czechoslovakian writer Karel Čapek wrote of a more scientific assembly of living matter:

. . . with the help of his tinctures he could make whatever he wanted. He could have produced a Medusa with the brain of a Socrates or a worm fifty yards long. But being without a grain of humor, he took it into his head to make a vertebrate or perhaps a man. This artificial living matter of his had a raging thirst for life. It didn't mind being sewn or mixed together. That couldn't be done with natural albumin. And that's how he set about it.

In Čapek's play *R.U.R.* (*Rossum's Universal Robots*) the scientist Rossum describes how he discovered this material from which he made his living robots. Off on some distant island to study the ocean fauna, he attempted a chemical synthesis of living matter. Instead he discovered a substance which behaved exactly like living matter, although its chemical composition was different. In Rossum's words: "Nature has found only one method of organizing living matter. There is, however, another method, more simple, flexible and rapid, which has not yet occurred to nature at all. This second process by which life can be developed was discovered by me today."

Rossum, the scientist, first produced an artificial dog, and then started on the "manufacture of a man." The result was mass production of worker robots. *R.U.R.* was a tract against industrialization, but Čapek also spelled out with clarity the basic procedures for factory production of life. His robots succeeded in killing off all but one human being in an uprising, stemming from their big flaw: they were sterile and couldn't produce more robots. Toward the end of the story they convince the lone human survivor to change them so they can reproduce. As two of the robots leave, he tells them, "Go, Adam. Go, Eve. The world is yours."

The Synthesis of Life

With an eye on Frankenstein and *R.U.R.,* some of us tend to picture mad scientists pouring smoking chemicals from beakers, then standing back while some horrible life form claws its way over the top. What is being done in biochemical laboratories engaged in research toward artificially producing or copying life is not nearly so titillating.

Evolutionists believe that life began in a primordial soup containing the ingredients of organic matter. Slowly, and purely by blind chance, these bits and pieces came together, stayed together, and gradually evolved into more complex assemblies. First came amino acids, then coacervates of proteins, and then cells. Cells began to divide and multiply, still entirely by chance and simply obeying physical and chemical laws that prevailed. All this blind activity finally led to human life—which is now reversing the process and trying to backtrack to where it all began.

Dr. Frankenstein, with the facility conferred by fiction, created a fairly recognizable "human being" right off the bat. Rossum cautiously produced a floppy dog before going to the top. In real life, the creation of life is far more difficult, and biologists attempting to emulate nature in the test tube begin much closer to the origin of life. The work of Dr. Harold Urey, Stanley Miller, and others is typical. Urey, a Nobel Prize winner, speculated that life originated much as Russia's A. I. Oparin guessed—with amino acids created as a result of proper constituents and physical and chemical conditions.

By 1963, Ceylonese biochemist Cyril Ponnamperuma synthesized the amino acid adenine in the laboratory. Cosmic rays were simulated by a high-energy beam of electrons, which linked up methane, ammonia, and water into several chemicals, including adenine, a constituent of DNA and RNA.

There were other important developments that year. Dr. Klaus Hofmann at the University of Pittsburgh split apart a molecule of ribonuclease, an enzyme with the property of breaking down nucleic acid in cells. Next he synthesized a substitute for the 20-amino-acid smaller portion of the ribonuclease. He used only 13 amino acids in his artificial model, but succeeded in joining it to the larger, living 104-unit portion. The hybrid ribonuclease molecule, even minus seven amino acids, proved to be about 70 percent as active as the original enzyme. Two other scientists, Drs. Waclaw Sybalski of Wisconsin and Rose Litman of Colorado, synthesized DNA itself in their labs. They also introduced it into bacteria and changed the hereditary characteristics of these hosts.

Leonard Engel wrote in 1962 that "cold-war-minded scientists have in fact urged a crash program to guarantee a U.S. first" in creating living matter in the laboratory. Three years later, Dr. Charles C. Price, then president of the American Chemical Society and head of the chemistry department at the University of Pennsylvania, proposed making the synthesis of life a national goal on a par with the man-on-the-moon project. So startling was this proposal to scientists and laymen alike that the American Chemical Society issued a quick disclaimer that Price's views were his own and did not necessarily reflect those of the Society.

Artificial life did not become an official "national goal," but progress has continued. In 1965 Dr. Sol Spigelman at the University of Illinois succeeded in synthesizing an RNA virus. This is probably the simplest living organism, for it consists mainly of genetic material, jacketed in an outer coating of protein. It possesses only one of the three normal properties of life: it can reproduce, or "replicate," itself by using raw materials from the host cell it invades. The virus synthesized by Spigelman and his colleagues was known as phi-Beta.

Although admittedly using material of biological origin in the form of an RNA primer, Spigelman's feat was hailed as "a delight for the materialist," since a living organism had been synthesized from synthetic precursors. The organism could then grow and multiply without further help from the biochemist who created it (with the help of a living pattern, of course).

By 1967, fourteen years after the first "life creation" experiments by Miller, a Florida researcher reportedly produced something like the coacervate assemblies theorized by Oparin in his book *Origin of Life*. Microspheres of proteinlike material resulted which even exhibited a budding division like that of bacteria to form chains of proteinlike material. Dr. Sidney Fox, director of the Institute of Molecular Evolution, University of Miami, produced these polymer assemblies, which he calls "protenoids."

Also in that year Dr. Arthur Kornberg and colleagues at Stanford University School of Medicine synthesized not just a constituent of DNA, but a length of such a molecule, in what was said to be an exact copy of nature. Indeed, President Lyndon B. Johnson lauded the scientists for their "awesome accomplishment," and Dr. James Shannon, director of the National Institutes of Health, stated that Kornberg had succeeded in what was "essentially the synthesis of life."

Other researchers felt that this was an exaggeration, since synthesis of a small amount of viral DNA hardly represented synthesis of life itself. Furthermore, living molecules controlled the synthesis of "off-the-shelf" chemicals Kornberg used to make up the artificial virus. Nevertheless, the synthesis of Phi X 174, a "pigmy virus," with a single strand of DNA joined in a circle, seemed a great step along the road toward creating life. In the spring of 1970, Har Gobind Khorana succeeded in synthesizing the double-stranded, 77-subunit molecule called yeast alanine transfer RNA. This is an entity that stands between the genes carrying hereditary information and the proteins whose composition the genes specify and produce.

The progress made by the life synthesizers is truly remarkable and could not have been foreseen two decades ago. DNA has been created in the test tube. DNA is the blueprint of life. Ergo, man is on the verge of creating artificial living things. However, with an artificial gene or even a complete and working nucleus, the synthesizer has only begun his work. The nucleus itself is but the control center; it must be installed in a proper environment to do its work of guiding and driving the force that converts raw materials of nature into living material. Making an artificial cell may be too tall an order, and the geneticist may have to settle for implanting an artificial set of chromosomes into an old-fashioned natural cell from which the original nucleus has been removed by radiation, chemistry, or other means. Even this feat, of course, would be the dawn of a new day in biology, since the proper genetic instructions can build from host cells. Given the genetic material, a proper cell environment, and nutrients, the rest should come easily, and artificial life will be a reality far beyond a mere stub of synthetic DNA or RNA.

Why Create Life?

More than a century after Frankenstein, his secret of creation had not been rediscovered. In *Brave New World,* Aldous Huxley had to use human cells to seed his assembly line that decanted fifteen types of people after the bottles containing them had traveled for 267 days—at 8 meters a day. Which suggests the questions in the minds of many, including some biologists: Why should we want to create life? In a world already threatened with too much *naturally* produced biomass, why should any scientist in his right mind try to produce an artificial competitor? If biologists should succeed in putting together off-the-shelf chemicals and creating an exact copy of life, would it not be the same as natural life if it is an exact copy?

With some 2 million different natural species already known, and thousands more being discovered each year, there would seem to be enough raw material in the world to obviate the need for factories turning out artificial life. The reason generally given by scientists—including Dr. Price, who suggested the crash program to achieve test-tube synthesis of life—is to gain knowledge useful elsewhere including the breeding of better plants and animals, the correction of disease, and other improvement of *natural* man.

It has been pointed out that in many cases nature not only provides perfectly adequate living material, it does it much cheaper. Take a human being, for example. The act of procreation generally does not cost anything. Although mothers "eat for two" this cost is not a large one. The biggest expense for a new baby is medical care, an item which primitive man did not find necessary and which may be charged to culture rather than any biological necessity. Although slavery is virtually eliminated, it is possible to acquire a human child for a relatively small sum, often merely for the taking. An artificial baby produced for the National Institutes of Health on a government grant might cost billions of dollars! On a smaller scale, it has been pointed out that adenine or alanine can be obtained more cheaply from natural sources than from laboratories which produce them artificially. At least that is the situation at the moment, and costs will probably not be reduced appreciably from the price of nature's pattern.

The average human brain is in most respects superior to a sophisticated third- or fourth-generation electronic computer costing hundreds of thousands of dollars. Yet the human brain can be mass-produced cheaply with unskilled labor, as has been pointed out by envious engineers. Perhaps there will be a future "organic computer," grown in a laboratory culture dish rather than assembled in an electronics plant, or a boat or an airplane grown rather than assembled or built. But these seem rather inglorious ways to use the miracle of artificial life.

Meantime, man has succeeded in creating a host of nonliving things like automobiles, airplanes, computers, and gas refineries that perform as living creatures cannot. Why endow such inorganic servants with the breath of life, and all the attendant problems that would bring?

The key, then, would seem to be the imagination of those designing the forms of artificial life to be produced in test tubes. Perhaps new pets can be fabricated. Perhaps new forms of life in the water and in the air can help solve the problems of pollution. Others may prove advantageous in space, and in planet exploration and colonization. The trick is to think of something *really* new, a living organism like none in existence today. Living rockets, or living liquid or gas, or vapor? A living building material that would *become* a house. A "live" entertainment form, a new "living" means of communication.

Immortality would seem a goal to shoot for in some applications, yet this quality might better be attained with something inorganic. The oldest living things seem to be the bristlecone pines of high mountain deserts, and these are only several thousand years old. Most living things give up the ghost before even a century has passed. On the other hand, bridges and dams endure for centuries, mountains for millions of years.

Despite the apparent uselessness of artificial men, however, it is possible that some natural men will pursue this goal, above board or illicitly, for good reason or bad. National pride is one goad; others are monetary profit, personal prestige, revenge, and so on down the scale of motivations. If a thing *can* be done, it may have a use and someone may be able to do it. If we fail to produce artificial life it most likely will not be for lack of trying.

Olaf Stapledon's *Last and First Men,* projecting the future of mankind as far into the future as anyone would care to consider, suggested four means of creating a superbrain—which is probably the kind of improvement artificial life creators would be after. Stapledon's four methods were: selective breeding, manipulation of genes in test-tube germ cells, manipulation of fertilized egg cells in the lab, and manipulation of the growing individual itself. None of these methods involved the creation of artificial life. The method favored by the scientists in Stapledon's story, a laboratory-fertilized natural human ovum, relied on the basic nucleus and cell provided by nature. The Creator has done such a marvelous job that perhaps there is no way man could improve on it as far as basics are concerned. Surely there are minor improvements to be made, but these do not entail designing a new life form from scratch.

For the first time in all time a living creature understands its
origin and can undertake to design its future.

Robert Sinsheimer, California Institute of Technology

13 The Genetic Prospect

At various stages in his evolutionary development, man has
stood at the threshold of a new era, a revolution in his life style.
The use of fire marked one of the first of these revolutions; others
came with speech and writing. Perhaps the greatest revolution was
agriculture, although some would say the industrial revolution
deserves that distinction. Since then there have been a number of
further revolutions—nuclear, computer, and so on. And now we
can discern what may be another crucial landmark in human development—the genetic revolution.

For a long time we have tampered with matter, and even with
nature itself. We have specialized our environment to the point
that most of us would be helpless if returned to primitive condi-

tions. These have been revolutionary changes, surely. But the genetic revolution is a different kind of change, for in it we are tampering with ourselves directly, poking (unskillfully as yet) at the genes that make us tick, in the hope of altering their measured beat. How well we carry through with this improvement program remains to be seen. But there is evidence that the genetic prospect includes potentially dangerous changes. In spite of this sobering knowledge, however, few of us seem to know—or care to know—much about what is taking place.

The Apathetic Patient

If the epigraph to this chapter from Dr. Sinsheimer (described as in the forefront of the genetic revolution) is true, it is certainly the best-kept secret since the atom bomb. This secrecy is the more remarkable because there has been no apparent attempt to hide what goes on in laboratories and hospitals around the world.

Robert Francouer is greatly troubled that even those whom he assumed would be concerned—Catholic readers, for example—do not respond at all; they simply fail to react to what is happening. Despite the earlier warnings of Jean Rostand, Aldous Huxley, and others, plus the concerns voiced presently, people generally refuse to consider such things seriously.

This lack of response is perhaps somehow allied to human attitudes about sex. It is remarkable that in an area so basically simple that one critic has said an intelligent baboon could adequately instruct its young in 15 minutes, human beings have managed to surround the reproductive act with a veil of ignorance, confusion, embarrassment, and misinformation. Even more remarkable is the fact that this squeamishness persists alongside what is hailed as an ongoing "sexual revolution."

Sex education in schools, fought tooth and nail by many parents, has often been made necessary by the refusal of those same mothers and fathers to instruct their own young in the facts of life, preferring to embroider fairy tales instead. Since genetic mechanics concerns the reproductive organs, there may be a mass refusal by society to admit that anything is going on. The average adult has surely been exposed to some basic Mendelian genetics, but most manage to hide that information away in a secret compartment of the brain, as if pretending it isn't there will guard against any embarrassing consequences.

It is not just among laymen that genetic diffidence exists. According to Nobel laureate James D. Watson, an analogous situation exists in the scientific community: "Though a number of scientific papers devoted to the problem of genetic engineering have casually mentioned that clonal reproduction may someday be with us, they have been so vague and devoid of meaningful time estimates as to be virtually soporific."

Perhaps part of the reason for such general apathy is that the dawn of a genetic revolution has been hailed periodically for a number of years, yet has never really lighted up the sky—at least not brilliantly enough to be noticed against a background of events of more immediate interest and easier-to-grasp significance.

Is It the Real Thing This Time?

Nobel Prize winner Alexis Carrel, who "immortalized" bits of chicken heart in his laboratory with Charles Lindbergh's help in the 1930s, wrote this in *Man, the Unknown:*

Science, which has transformed the material world, gives man the power of transforming himself. It has unveiled some of the secret mechanisms of his life. It has shown him how to alter their motion, how to mold his body and his soul on patterns born of his wishes. For the first time in history, humanity, helped by science, has become master of its destiny. But will we be capable of using this knowledge of ourselves to our real advantage? To progress again, man must remake himself. And he cannot remake himself without suffering. For he is both the marble and the sculptor. In order to uncover his true visage he must shatter his own substance with heavy blows of his hammer. . . .

There were 55 printings of this book, but perhaps readers of that generation became disenchanted when no genetic saviors appeared to sculpt man in a better image.

In 1938 Lancelot Hogben announced in *Science for the Citizen:*

Evolution unfolds a new horizon of human destiny. Man has it in his power to become an active and intelligent directive agent in the evolutionary process, using his knowledge of the diversity of living creatures to decide which are essential to his own welfare as objects of use or of esthetic satisfaction, and using his knowledge of the properties of living matter to adjust the environment of the species he chooses as members of a rationally planned ecological system. The bio-technical future of man is not limited to these two themes. We also have it in our power to set about creating new types of organisms—and perhaps ultimately of guiding the further evolution of the unborn capacities of our own species.

Part of the reason this news fell on deaf ears may have been that "ecology" was awaiting discovery by more popular writers. At any rate, in the thirty-five years since Hogben heralded man's Promethean powers, the millennium comes only creepingly to humanity. Progress, surely, but not the sweeping kind promised by these far-sighted commentators.

The Irresistible Force

Regardless of the attitudes of laymen or scientists, the glacier is beginning to change its course at long last, however. Man is more and more a conscious factor in shaping his own future. We now know that atomic energy has a side effect that may not be good. We know that pollution is an environmental problem with dangerous implications for both the present and the future. But serious as these factors are, they do not equal those of genetic engineering.

In 1971 the International Atomic Energy Agency sponsored a conference in Vienna on "The Use of Genetic Improvement of Industrial Microorganisms." About a hundred participants representing twenty-seven countries and six regional or international organizations attended the meeting, whose purpose was to spread the word on industrial uses of genetic mutations. Focus was on mutagenic agents that permit gene changes toward beneficial alterations in fermentation and other processes. Very interesting work was described, including the discovery of minibacteria having no nuclei and lacking the ability to reproduce. Giant cells, able to grow 500 to 1,000 times as large as normal cells, were reported. Scientists have also produced "minicells."

While the conference was industry-oriented, medical implications were also discussed. If beneficial mutations are important in waste removal and industrial processes, surely they are important medically. If a geneticist can mix the DNA from one cell with that of a different cell for industry, he can do it for animals and man as well. For various reasons, he *will* do it, as James D. Watson points out. And, according to him, herein lies the problem:

This is a matter far too important to be left solely in the hands of the scientific and medical communities. The belief that surrogate mothers and clonal babies are inevitable because science always moves forward, an attitude expressed to me recently by a scientific colleague, represents a form of laissez-faire nonsense dismally reminiscent of the creed that American business, if left to itself, will solve everybody's problems.

Do We or Don't We?

Robert Sinsheimer, almost in the next breath after announcing our new power over our future, questions the wisdom of using that power and asks: "For what purpose should we alter our genes?" Another Californian, Sam Goldwyn, is often misquoted to the effect that all one need do to acquire writing style is nothing. Could it also be that all we need do to attain our evolutionary destiny is nothing?

In the book *Man and His Future,* edited by Gordon Wolstenholme, the following observation is made:

As Darwin pointed out, there has been during biological evolution a general trend towards improvement—improvement in efficiency and self-regulation. This trend is inevitable, but is accompanied by much waste, suffering and extinction. The trend towards improvement continues in psycho-social evolution, although again accompanied by suffering, horror and evil. Yet in spite of all the waste and misery, the total improvement achieved during the whole process of evolution from the origin of life to the present day is almost incredible—that from a sub-microscopic precellular colloid to a self-conscious, civilized, human vertebrate throwing up on its way a fantastic profusion of organic and cultural variety.

Both these views seem to make a case against genetic engineering. If God is in His heaven, it is presumptuous of us to try to improve on His handiwork. If blind chance has done all this, why exert ourselves instead of just sitting back and enjoying the ride? Yet the die seems to be cast, and humanity about to cross the river into a land whose promise we can only guess.

Hermann Muller worried until his death about the coming "genetic eclipse" brought on by the increasing genetic load. Geneticist Theodosius Dobzhansky currently points out the ethical dilemma facing us: helping the weak and the deformed to live and reproduce their kind leads to the prospect of a *genetic twilight,* he admits, but if we let them die or suffer when we can help them, we face the certainty of a *moral twilight.* We seem to be damned if we do and doubly damned if we don't.

The crucial question is the matter of degree to which the camel follows his nose into the tent. Having eliminated negatives like sickle-cell anemia, dyslexia, cystic fibrosis, cancer, and more, will we then engineer our genes for "positive" change, as in Huxley's Central Hatchery? It does seem, because of population growth and other problems, that major adjustments are in prospect for society in the future. However, some planners still see no need for employing genetic engineers on what is basically a job for behavioral psychologists. According to B. F. Skinner, in *Beyond Freedom and*

Dignity, "The problem is to design a world which will be liked not by people as they now are but by those who live in it." The key here is the shape of the outside world, not that of the inner man. There is no need for a geneticist to resuscitate mankind, for all the world like a lifeguard as he chants to himself, "Out go the *bad* genes, in come the *good* genes! Out go the *bad* genes. . . ."

Skinner does express some wonderment over the failure of behavioral psychology to remake the world before now. Many others see the whole behaviorist approach as a dismal failure and urge a belated facing of the genetic facts of life. This "radical" idea has been voiced down through the ages, from Plato to Galton, and was echoed ringingly in *Brave New World* in 1932.

Huxley's social planners were aiming at the same world as Skinner's: "And that is the secret of happiness and virtue—liking what you've got to do. All conditioning aims at that. Making people like their unescapable destiny." But they had to turn to genetic engineering to get the job done. "In the end, the Controllers realized that force was no good. The slower but infinitely surer methods of ectogenesis, neo-Pavlovian conditioning and hypnopaedia. . . ."

An engineered society will require all kinds of people, of course, shaped differently to fit into different societal slots and niches:

"I was wondering," said the Savage, "why you had them [Epsilon Semi-Morons] at all—seeing that you can get whatever you want out of those bottles. Why don't you make everybody an Alpha Double Plus while you're about it?"

Mustapha Mond laughed. "Because we have no wish to have our throats cut," he answered. "We believe in happiness and stability. A society of Alphas couldn't fail to be unstable and miserable. Imagine a factory staffed by Alphas—that is to say by separate and unrelated individuals of good heredity and conditioned so as to be capable (within limits) of making a free choice and assuming responsibilities. Imagine it!"

There are other fears of what might be done with genetic control, fears of a wilder nature that recall science-fiction scare movies on the subject. Geneticist Salvador Luria even discusses genetic warfare. He suggests that unscrupulous persons might discover a virus that could cause a tremendous sensitivity to carbon dioxide in people. This genetic trait apparently exists in fruit flies, and Luria says that someone might be tempted to gain control over humanity by spreading such a virus.

There are also fears of such "genetic weapons" as contraceptive agents added to the public water supply to enforce population control, and mandatory abortion or sterilization for women who have had "enough" children or who may not produce the "proper" kind. In response to those who cannot believe in such inhumanity, the concerned remind scoffers of purges and liquidations of millions of undesirables carried out by Hitler and in Communist countries.

Who Holds the Genetic Strings?

In questioning the use of genetic engineering, Dr. Sinsheimer also asks, "To whom should we give such powers?" Who will do the controlling, who will make the strategic decisions of abortion, sterilization, gene therapy, ectogenesis, germinal choice, and other genetic manipulations? Is this the province of individuals, to whom future politicians will promise: "The genes of your own choice!" or will genetic choice reside with society in the larger sense—or even solely with the rulers?

Alexis Carrel suggested a scientific branch of government set up to consider future developments, particularly those in genetics. This idea has gathered force. Earlier we noted the activities of the Institute of Society, Ethics, and the Life Sciences in monitoring developments in genetics. The Texas Medical Center in Houston has a similar organization called the Institute of Religion and Human Development at the Baylor College of Medicine. Cancer researcher Van Rensselaer Potter, of the University of Wisconsin, has proposed in his book *Bioethics* that the government create a fourth branch which Potter calls a "Counsel for the Future."

James D. Watson is so fearful of uncontrolled research that he recently appeared before a congressional committee to urge a ban on all human-cell experiments toward unnatural methods of reproduction, unless international agreements on research limitations can be reached and enforced. At least one scientist has taken drastic personal action: Dr. James Shapiro, a biologist working on bacterial genes, reportedly changed his field to social work for fear of misuse of genetics. On the other hand, some called this a "failure of nerve" on Shapiro's part.

Passage of the Mondale Bill, which would provide for analysis of new scientific problems, has been urged, and with all the growing stress on "technology assessment" it seems inevitable that the government will become increasingly involved. To be sure, this eventuality gives little comfort to some, but alternatives are difficult to arrive at.

The United Nations Council for International Organizations of Medical Sciences, a part of the World Health Organization and UNESCO, late in 1972 proposed the creation of an international nongovernmental body to explore the moral and social issues raised by the developments in genetics. Amitai Etzioni, whose concern over sex selection of children was described earlier, suggests the approaches such an organization might consider: amniocentesis might be recommended for all women over forty, living in countries permitting abortion, to guard against mongolism. Guidelines might also be necessary to prevent abuses in the selection of the sex of a child.

Homo Futurus

The major revolutions in man's earlier history all were similar in one respect: they resulted in population explosions. Ironically the genetic revolution may be the first to lead to a curtailment of the number of human beings. Although it is questionable that all of us are agreed on the need for population control, it is a fact that birth control is now far more effective than ever before in our history. It has been suggested that the genetic revolution may produce a new human species, one that might well be called *Homo futurus*. The question we should ask is whether such a new breed would flourish in quantity or in quality. Or whether it might not survive at all but rather pass away as 98 million earlier species of life have already done.

Bibliography

American Association for the Advancement of Science, *Advances in Human Genetics and Their Impact on Society*. New York: The National Foundation of the March of Dimes, 1972.

Asimov, Isaac. *The Genetic Code*. New York: Signet Books, 1963.

Augenstein, Leroy. *Come, Let Us Play God*. New York: Harper & Row, 1969.

Calvin, Melvin. *Chemical Evolution*. New York: Oxford University Press, 1969.

Carrel, Alexis. *Man, the Unknown*. New York: Harper & Row, 1939.

Cavalli-Sforza, L. L., and Bodmer, W. F. *The Genetics of Human Populations*. San Francisco: W. H. Freeman Co., 1971.

Comfort, Alex. *The Process of Ageing*. New York: New American Library, 1964.

Darwin, Charles. *The Origin of Species*. New York: Washington Square Press, 1963.

Davis, F. B. *The Measurement of Mental Capacity Through Evoked Potentials.* Greenwich, Conn.: Educational Records Bureau, 1971.

Dobzhansky, Theodosius. *Heredity and the Nature of Man.* New York: Harcourt, Brace & World, 1964.

Eysenck, H. J. *The IQ Argument.* New York: Library Press, 1971.

Francouer, Robert T. *Evolving World, Converging Man.* New York: Holt, Rinehart & Winston, 1970.

————. *Utopian Motherhood.* New York: A. S. Barnes, 1972.

Galton, Francis. *Hereditary Genius.* New York: World Publishing Co., 1962.

Guilford, J. P. *The Nature of Human Intelligence.* New York: McGraw-Hill, 1967.

Halacy, D. S., Jr. *Cyborg: Evolution of the Superman.* New York: Harper & Row, 1965.

Handler, Philip (ed.). *Biology and the Future of Man.* New York: Oxford University Press, 1970.

Harris, Maureen (ed.). *Early Diagnosis of Human Genetic Defects.* Washington, D.C.: U.S. Government Printing Office, 1972.

Harvard Educational Review. *Environment, Heredity, and Intelligence.* Cambridge, Mass.: Cambridge University Press, 1969.

Hogben, Lancelot. *Science for the Citizen.* New York: Alfred A. Knopf, 1938.

Huxley, Aldous. *Brave New World.* New York: Harper & Row, 1946.

Jencks, Christopher. *Inequality: A Reassessment of the Effect of Family and Schools in America.* New York: Basic Books, 1972.

Lang, Theo. *The Difference Between a Man and a Woman.* New York: John Day, 1971.

Mayr, Ernst. *Populations, Species, and Evolution.* Cambridge, Mass.: Belknap Press, 1970.

Medvedev, Zhores A. *The Rise and Fall of T. D. Lysenko.* New York: Columbia University Press, 1969.

Monod, Jacques. *Chance and Necessity.* New York: Random House, 1971.

Morgan, Elaine. *The Descent of Woman.* New York: Stein & Day, 1972.

Muller, Hermann J. *Studies in Genetics: The Selected Papers of H. J. Muller.* Bloomington: Indiana University Press, 1962.

Oparin, A. I. *The Origin of Life.* New York: Dover Publications, 1953.

Osborn, Frederick. *The Future of Human Heredity: An Introduction to Eugenics in Modern Society.* New York: Weybright and Talley, 1968.

Potter, Van Rensselaer. *Bioethics: Bridge to the Future.* Englewood Cliffs, N.J.: Prentice-Hall, 1971.

Ramsey, Paul. *Fabricated Man: The Ethics of Genetic Control.* New Haven: Yale University Press, 1910.

Rosenfeld, Albert. *The Second Genesis: The Coming Control of Life.* Englewood Cliffs, N.J.: Prentice-Hall, 1968.

Rostand, Jean. *Can Man Be Modified?* New York: Basic Books, 1950.

Shettles, Landrum B., and Rorvik, David. *Your Baby's Sex; Now You Can Choose.* New York: Dodd, Mead, 1970.

Sonneborn, Tracy M. *The Control of Human Heredity and Evolution.* New York: Macmillan, 1965.

Taylor, Gordon Rattray. *The Biological Time Bomb.* New York: World Publishing Co., 1968.

Thielicke, Helmut. *The Ethics of Sex.* New York: Harper & Row, 1964.

Wolstenholme, Gordon E. W. (ed.). *Man and His Future.* Boston: Little, Brown, 1963.

Young, Louise B. (ed.). *Evolution of Man.* New York: Oxford University Press, 1970.

Index